ALLUVIAL FAN FLOODING

Committee on Alluvial Fan Flooding

Water Science and Technology Board

Commission on Geosciences, Environment, and Resources

National Research Council

NATIONAL ACADEMY PRESS
Washington, D.C.   1996

NOTICE: The project that is the subject of this report was approved by the Governing Board of the National Research Council, whose members are drawn from the councils of the National Academy of Sciences, the National Academy of Engineering, and the Institute of Medicine. The members of the committee responsible for the report were chosen for their special competences and with regard for appropriate balance.

This report has been reviewed by a group other than the authors according to procedures approved by a Report Review Committee consisting of members of the National Academy of Sciences, the National Academy of Engineering, and the Institute of Medicine.

Support for this project was provided by the U.S. Federal Emergency Management Agency under Contract Agreement EMW-94-C-4550.

Library of Congress Catalog Card Number 96-69351
International Standard Book Number 0-309-05542-3

Additional copies of this report are available from:

National Academy Press
2101 Constitution Ave., NW
Box 285
Washington, DC 20055
800-624-6242
202-334-3313 (in the Washington Metropolitan Area)
http://www.nap.edu

Cover credit: Alluvial fan flooding at Magnesia Spring Canyon in July 1979 caused one death and more than $7 million in damage. Photograph taken from 1989 FEMA Document 165, *Alluvial Fans: Hazards and Management.*

COMMITTEE ON ALLUVIAL FAN FLOODING

STANLEY A. SCHUMM, *Chair*, Colorado State University, Fort Collins, Colorado
VICTOR R. BAKER, University of Arizona, Tucson
MARGARET (PEGGY) F. BOWKER, Nimbus Engineers, Reno, Nevada
JOSEPH R. DIXON, U.S. Army Corps of Engineers, Phoenix, Arizona
THOMAS DUNNE, University of California, Santa Barbara
DOUGLAS HAMILTON, engineering consultant, Irvine, California
HJALMAR W. HJALMARSON, consultant, Camp Verde, Arizona
DOROTHY MERRITTS, Franklin & Marshall College, Lancaster, Pennsylvania

Staff

CHRIS ELFRING, Study Director
ANGELA BRUBAKER, Research Assistant
ETAN GUMERMAN, Research Associate
ROSEANNE PRICE, Consulting Editor

The National Academy of Sciences is a private, nonprofit, self-perpetuating society of distinguished scholars engaged in scientific and engineering research, dedicated to the furtherance of science and technology and to their use for the general welfare. Upon the authority of the charter granted to it by the Congress in 1863, the Academy has a mandate that requires it to advise the federal government on scientific and technical matters. Dr. Bruce M. Alberts is president of the National Academy of Sciences.

The National Academy of Engineering was established in 1964, under the charter of the National Academy of Sciences, as a parallel organization of outstanding engineers. It is autonomous in its administration and in the selection of its members, sharing with the National Academy of Sciences the responsibility for advising the federal government. The National Academy of Engineering also sponsors engineering programs aimed at meeting national needs, encourages education and research, and recognizes the superior achievements of engineers. Dr. William A. Wulf is interim president of the National Academy of Engineering.

The Institute of Medicine was established in 1970 by the National Academy of Sciences to secure the services of eminent members of appropriate professions in the examination of policy matters pertaining to the health of the public. The Institute acts under the responsibility given to the National Academy of Sciences by its congressional charter to be an adviser to the federal government and, upon its own initiative, to identify issues of medical care, research, and education. Dr. Kenneth I. Shine is president of the Institute of Medicine.

The National Research Council was organized by the National Academy of Sciences in 1916 to associate the broad community of science and technology with the Academy's purposes of furthering knowledge and advising the federal government. Functioning in accordance with general policies determined by the Academy, the Council has become the principal operating agency of both the National Academy of Sciences and the National Academy of Engineering in providing services to the government, the public, and the scientific and engineering communities. The Council is administered jointly by both Academies and the Institute of Medicine. Dr. Bruce M. Alberts and Dr. William A. Wulf are chairman and interim vice chairman, respectively, of the National Research Council.

Preface

People have long elected to build in flood-prone areas—whether because they sought easy access to the waterways that were once our main transportation routes, because they offer relatively flat building sites, or because of their aesthetic appeal. As the population increases and people search for desirable locations to live, they sometimes come into conflict with those who regulate construction on floodplains. In the western United States, some of the most intense conflicts revolve around development on alluvial fans, which can be susceptible to a particularly catastrophic type of flooding. Controversy over alluvial fan flooding issues led the Federal Emergency Management Agency (FEMA) to ask the National Research Council (NRC) for help. As a result, the NRC established the Committee on Alluvial Fan Flooding with a membership composed of eight engineers and earth scientists, all of whom have experience with alluvial fan morphology and processes.

The committee was charged to revise the existing definition of alluvial fan flooding, to develop criteria to determine if an area is subject to alluvial fan flooding, and to provide examples of the application of the definition and the criteria used. The committee recognized immediately that in addition to "alluvial fan flooding," there exists a broader category termed "uncertain flowpath flooding" that requires further consideration by FEMA. Confusion caused by linking two aspects of the flood hazard (i.e., land form type and uncertainty in flood processes) is part of the reason for the controversy on this subject. This committee cannot claim to have the final word on what it considers to be a complex technical and regulatory issue, hence we may not have achieved everything desired by FEMA. It has, however, provided significant guidance for characterizing how floods occur on alluvial fans and describing how FEMA might more consistently administer the National Flood Insurance Program on such land forms, which comprise large areas of the western United States and elsewhere.

In order to more fully understand the problems associated with alluvial fan flooding, the committee met at three locations in Arizona, California, and Utah, where different alluvial fans could be visited and evaluated in the field. The examples ranged from typical large alluvial fans in Arizona and California to small debris flow fans in Utah. Fans ranged from fully active, where flooding or debris flows could occur anywhere on the fan, to incised, where the bulk of the fan is not subject to flooding. Hence, not only was the varied expertise of the committee brought to bear on the problem, but the members were exposed in the field to new and different situations.

The committee benefited greatly from presentations and guidance in the field from the following people: Gary Christiansen and Mike Lowe, Utah Geological Survey; Fred Campbell, EIS Engineering, Salt Lake City; Sidney Smith, Davis County Public Works, Utah; Jeffrey Keaton, AGRA Earth and Environment, Salt Lake City; Joseph Tram, Maricopa County Flood Control District, Arizona; Terri Miller, Arizona Department of Water Resources; Philip Pearthree, Arizona Geological Survey; Joseph Hill, San Diego County Department of Public Works; Stuart McKibbin, Riverside County, California; Robert Mussetter, Mussetter Engineering, Fort Collins, Colorado; James Slosson, Slosson and Associates, California; and Joe Cook, Coachella Valley Water District. We also appreciate the support provided by FEMA personnel and contractors—especially Frank Tsai, Karl Mohr, and Ed Mifflin—who helped us understand the issues and how FEMA currently operates. We believe the hands-on perspective that all these people contributed was essential to the evolution of our thinking. In addition, the committee would like to thank the staff of the Water Science and Technology Board for their invaluable guidance to the committee, especially the insights provided by study director Chris Elfring and support from her associates Angela Brubaker and Etan Gumerman. Our thanks also to Tamera Benson for the preparation of the graphics.

Stanley Schumm, *Chair*
Committee on Alluvial Fan Flooding

Contents

Summary

Alluvial fans are gently sloping, fan-shaped landforms created over time by deposition of eroded sediment, and they are common at the base of mountain ranges in arid and semiarid regions such as the American West. Given that alluvial fans tend to occur in apparently dry conditions, homeowners are often shocked to find that they can be the sites of destructive floods. Floods on alluvial fans, although characterized by relatively shallow depths, can strike with little warning, can travel at extremely high speeds, and can carry tremendous amounts of sediment and debris. Such flooding presents unique problems to federal and state planners in terms of quantifying the flood hazards, estimating the magnitude at which those hazards can be expected at a particular location, and devising reliable mitigation strategies.

The Federal Emergency Management Agency (FEMA) has great influence over the way communities manage and mitigate flood hazards. FEMA's influence comes both from its congressional mandate and from its role as enforcer of National Flood Insurance Program (NFIP) regulations. When FEMA designates an area as subject to alluvial fan flooding, rather than ordinary riverine flooding, it sets in motion specific, restrictive federal regulations. Because such a designation can affect development opportunities, it can be controversial.

NFIP regulations define alluvial fan flooding to be "flooding occurring on the surface of an alluvial fan or similar landform which originates at the apex and is characterized by high velocity flows; active processes of erosion, sediment transport and deposition; and unpredictable flow paths." In addition, although *alluvial fan flooding* is a general term that can involve flooding over an entire surface, the FEMA mandate is to determine the extent of hazard associated with a flood with a 100-year recurrence interval (i.e., a 1 percent probability in a given year). Hence, the term *alluvial fan flooding* is used in two ways. In the geomorphic sense, it can be any flood on an alluvial fan. But in the NFIP sense, it is the distribution of 100-year floodwater on the fan. The reader is cautioned that the term is used in both ways, including in this report.

The problem with the current definition is that it is very broad, and often is applied to many landforms that are not alluvial fans, such as alluvial plains, pediments, deltas, and braided streams. One approach to reduce this confusion is to define alluvial fan flooding so that it applies strictly to alluvial fans, and to use different language, such as uncertain flow path flooding, when dealing with "similar landforms." But such a change is not as simple as it may sound—it requires agreement on the definition of alluvial fan flooding and clear guidelines that can help planners, regulators, and citizens reach a common understanding of what an alluvial fan is and when it presents a flood hazard. To help FEMA with this problem, the Committee on Alluvial Fan Flooding was established and charged to develop a revised definition of alluvial fan flooding, to specify criteria that can be assessed to determine if an area is subject to alluvial fan flooding, and to provide examples that illustrate the definition and criteria.

To begin, the committee needed a clear definition of "alluvial fan." Working from standard geologic definitions, the committee defines an alluvial fan to be "a sedimentary deposit located at a topographic break, such as the base of a mountain, escarpment, or valley side, that is composed of streamflow and/or debris flow sediments and that has the shape of a fan either fully or partially extended." This deposit is convex in cross-profile. On a smooth cone-shaped fan, floodwater can spread widely across the surface in the same way that marbles will follow random paths down a gently sloped surface. Alluvial fans evolve through geologic time, and their evolution is affected by climate change and tectonics, and therefore a wide variety of fan morphologies can be observed, from the ideal smooth surface on which flow paths can be predicted only with great uncertainty to deeply incised fans with flow confined to a single channel. In the latter case, the flow path can be predictable, and the fan surface is not susceptible to major flooding. As a result, neither the automatic assumption of uniform flood risk on an alluvial fan nor the acceptance of complete uncertainty of flooding across an alluvial fan is reasonable.

The committee decided that the first step necessary to reduce the confusion was to define alluvial fan flooding as a flood hazard that occurs only and specifically on alluvial fans. According to the committee, alluvial fan flooding is characterized by flow path uncertainty so great that this uncertainty cannot be set aside in realistic assessments of flood risk or in the reliable mitigation of the hazard. The committee has determined that an alluvial fan flooding hazard is indicated by three related criteria: (1) flow path uncertainty below the hydrographic apex, (2) abrupt deposition and ensuing erosion of sediment as a stream or debris flow loses its competence to carry material eroded from a steeper, upstream source area, and (3) an environment where the combination of sediment availability, slope, and topography creates an ultrahazardous condition for which elevation on fill will not reliably mitigate the risk.

The committee notes that alluvial fan flooding typically begins to occur at the hydrographic apex, which is the highest point where flow is last confined, and then spreads out as sheetflood, debris slurries, or in multiple channels along paths that are uncertain. The hydrographic apex may be at or downstream of the topographic apex. Such flooding is characterized by sufficient energy to carry coarse sediment at shallow flow depths. The abrupt deposition of this sediment or debris strongly influences hydraulic conditions during the event and may allow higher flows to initiate new, distinct flow paths of uncertain direction. Also, erosion strongly influences hydraulic conditions when flood flows enlarge the area subject to flooding by undermining channel banks or eroding new paths across the unconsolidated sediments of the alluvial fan. Flow path uncertainty on the fan is aggravated by the absence of topographic confinement or by the occurrence of erosion and deposition. Flow path uncertainty at the hydrographic apex can be aggravated by deposition early in the flood that results in overbank flooding from a channel that otherwise appears too large to overflow. Such channel filling can be eroded during later stages of the flood. Together, these characteristics create a flood hazard that can be reliably mitigated only by the use of major structural flood control measures that require careful maintenance or by complete avoidance of the affected area.

An alluvial fan is a sedimentary deposit located at a topographic break that is composed of fluvial and/or debris flow sediments and that has the shape of a fan either fully or partially extended, as illustrated by the Hanaupah Canyon alluvial fan in Death Valley, California. Courtesy of H. W. Hjalmarson.

The committee also notes that the potential for erosion and deposition, the related uncertainty in flow path behavior, and the imprudence of elevation on fill as a mitigation measure are joint and separate characteristics shared among many flood hazards on depositional environments other than alluvial fans, although not usually with the same intensity. It stands to reason that some of the same rules should apply to this more inclusive type of flood hazard, which the committee calls *uncertain flow path flooding*, as apply to alluvial fan flooding, which is, in fact, a type of uncertain flow path flooding.

In the simplest case, a fan is shaped like a simple cone emanating from a single, well-defined apex. In such a case, a stream follows more-or-less radial paths down the cone, and the contours on the map of such a fan are convex downslope. However, the fan shape may not always be so apparent; for instance, it is obscured where the sedimentary accumulations from several source areas encroach on one another. At their downstream margins, fans merge with smoother depositional topography of the valley floor, river terraces, and lake and coastal deposits, and the channels may be small, shallow, and diffuse. Fans differ from pediments, some of which are cone-shaped, in that fans are formed by the accumulation of sediment, while pediments are erosional surfaces that are usually covered by a thin veneer of alluvium and colluvium.

Although alluvial fans are often thought to occur mainly in the western United States, they occur in a wide range of environments, including the Appalachian Mountains, western Canada, and various montaine, arid, and volcanic regions around the world. In North America, most fans that are subject to controversy are in the West because it is a rapidly urbanizing region and fans—with their relatively gentle terrain and views of the mountains—are appealing building sites.

Alluvial fans, and alluvial fan flooding, show great diversity because of variations in climate, fan history, rates and styles of tectonism, source area lithology, vegetation, and land use. For this reason, it is essential that any investigation of alluvial fan flooding include careful examination of the specific fan for which information is needed by specialists experienced in the study of alluvial fan processes and recognition of geomorphic indications of past and present flooding. The committee recognizes that the extent of site-specific examination may be constrained by factors such as the amount of time and money allocated to the project, the tools available to the investigator, and the investigator's experience. Nevertheless, it is essential to conduct at least one field inspection of every fan being delineated—to walk across its surfaces and along its channels.

The criteria used to assess whether an area is, or is not, subject to alluvial fan flooding must help the observer determine first, whether the area is a fan, and second, whether it is characterized by sedimentation and flow path uncertainty. Thus the process of determining whether or not an area is subject to alluvial fan flooding, and of defining the spatial extent of such flooding, can be divided into three stages:

1. Recognizing and characterizing alluvial fan landforms.
2. Defining the nature of the alluvial fan environment and identifying active and inactive components of the fan; and
3. Defining and characterizing areas of the fan affected by the 100-year flood.

Progression through each of these stages results in a phased procedure that narrows the problem to smaller and smaller areas. In Stage 1, the landform on which flooding occurs must be characterized. If the location of interest is an alluvial fan, then the user progresses to Stage 2, in which those parts of the alluvial fan that still are active are identified. The term *active* means that flooding, deposition, and erosion have occurred on the fan and might continue to occur on that part of the fan. Those parts of the fan that have been active in recent time can be identified depending on data availability for the site and money allocated to the project. Each active part of the alluvial fan also is characterized based on the dominant types of processes that result in sedimentation. Finally, in Stage 3 the user determines whether or not flooding by the 100-year flood is still probable on those parts of the fan that still are active and estimates the extent of such flooding. Progression through these stages will require a variety of maps and photos, as well as a significant amount of fieldwork and analysis to fully understand the flood hazard.

The effects of erosion and deposition processes and flow path uncertainty on flood hazard severity are not limited to alluvial fans. Yet the term *alluvial fan flooding* suggests these processes are limited to alluvial fans and is therefore confusing. The Committee on Alluvial Fan Flooding recommends that the term *alluvial fan flooding* be applied only to flooding on alluvial fans. FEMA will need to develop a strategy to regulate other types of uncertain flow path flooding that do not occur on alluvial fans.

This report addresses a wide range of issues related to alluvial fan flooding. Chapter 1 presents an introduction to why identification of alluvial fan flooding hazards is controversial and the problems of definitions. Chapter 2 looks in more depth at fan types and flooding processes. Chapter 3 presents indicators developed to help delineate alluvial fans and alluvial fan flooding, based on the committee's definition and discusses methodologies to delineate flood hazards on alluvial fans. Chapter 4 contains seven examples analyzed by the committee in light of the definition and field criteria. The sites represent a range of flood processes, from unconfined water flooding and debris flows on untrenched active fans to confined water flooding in fully trenched inactive alluvial fans; the examples also show variable amounts of study—from intensive to casual. Chapter 5 presents a summary of the committee's conclusions and recommendations.

Key conclusions include the following:

- Site investigation is essential to distinguish alluvial fans from other landforms and to identify which parts of an alluvial fan are subject to hazard.
- Regulatory flexibility is necessary to realistically depict flood hazards given the variability in flood processes on alluvial fans.
- The existing regulatory framework, which divides all flooding sources into either riverine or alluvial fan flooding, leads to inconsistency when imposed on specific sites.
- Imposing the alluvial fan flooding paradigm instead of the riverine paradigm creates its own set of difficulties for sound regulation of the flood hazard.
- The act of defining the type of flooding is independent from the act of deciding which methods are applicable for delineating the boundaries of the hazard.
- The role of uncertainty in mapping flood hazards on alluvial fans is different from that for floodplain management and mitigation.

Key recommendations include the following:

- The existing NFIP definition of alluvial fan flooding should be revised to reduce confusion and controversy. As noted earlier, this committee proposes a definition that limits the term to use only on alluvial fans and for FEMA purposes to the 100-year flood.
- FEMA also can recognize that uncertain flowpath flooding includes alluvial fan flooding as well as flooding on alluvial plains, deltas, and other landforms on which flowpaths change.
- During the delineation process, site-specific evaluation must be conducted because it is the key to determining which alluvial fans and parts of alluvial fans are subject to flood hazards.
- When estimating flood hazards, FEMA should evaluate uncertainty directly instead of assuming it to be either nonexistent or random.
- FEMA needs to expand the technical and regulatory input it receives in the delineation and regulation process, perhaps through the use of a technical advisory board composed of earth scientists, engineers, local regulating bodies, and those being regulated.
- If FEMA elects to extend the current alluvial fan regulatory construct to any nonalluvial fan situation, it will need to change the term *alluvial fan flooding* to *uncertain flow path flooding*.

1

Introduction

The Federal Emergency Management Agency (FEMA), which administers the National Flood Insurance Program (NFIP), is authorized to identify natural hazards throughout the United States and its territories. The geographical diversity of the nation provides a wide range of natural hazards, but one of FEMA's key responsibilities is to map areas that are subject to a 1 percent probability of being flooded in any year (the "100-year flood"). The purpose of this charge is to meet the NFIP's requirement that the burden of paying for flood damage be shifted from the general public to those living at risk. In most riverine environments, where channels change their locations only gradually and where catastrophic alterations in their form and flood conveyance capacity during a single event are rare, the procedures for mapping the depth and velocity of floods are generally agreed on. The technical and regulatory community has developed certain language, procedures, and a way of depicting reality (i.e., a paradigm) that allows the identification, delineation, and mitigation of flood hazards (see, e.g., Hydrologic Engineering Center, 1976, volume 6; and Bedient and Huber, 1992, Chapter 7). Although all floods behave, in detail, differently from the paradigm, once an estimate of the 1 percent peak flood discharge is agreed on, institutionalized procedures make the calculation of the extent, depth, and velocity of the flood hazard relatively straightforward and reproducible by different analysts. This report uses the term *riverine flooding* to represent those cases where application of this standard paradigm allows one to successfully assess and manage flood risk.

However, where catastrophic changes in river channel form and position can occur during a single flood, the traditional paradigm and associated hydraulic procedures cannot be relied on. For example, if a flood deposits large quantities of sediment on the channel bed in a reach, the conveyance capacity of the channel could be reduced drastically and the flow forced overbank at a lower discharge than would be predicted from prestorm surveys of the channel geometry. If overbank flooding causes erosion of a new channel or the reoccupation of an old channel, flood risk assessments based on the historical flow path would misrepresent the location and intensity of flooding downstream of the change. Both of these types of channel changes (form and position) can occur with great frequency and intensity on a type of landform called an alluvial fan. An alluvial fan, as defined by this committee, is "*a sedimentary deposit located at a topographic break, such as the base of a mountain front, escarpment, or valley side, that is composed of fluvial and/or debris flow sediments and which has the shape of a fan either fully or partially extended."*

Despite the fact that to geologists *alluvial* refers strictly to features and materials deposited by streams of water (American Geological Institute, 1976), in the case of fans the term has been used more loosely in the scientific, engineering, and planning literature to refer to the products of streams or debris flows. This usage is too well established for this committee to reverse. However, to a person actually analyzing flood risk on the ground, the distinction is important, and so the committee will recognize the needs of these people in Chapter 2, which is concerned with the physical processes associated with flooding rather than policy and regulatory aspects. Chapter 2 explains that alluvial fans can be deposited by streams or by debris flows or by some combination of the two, and that recognition of the difference between these situations can be crucial for correctly identifying the hazard potential in certain areas. The chapter thus distinguishes between stream-flow fans, debris flow fans, and composite fans. However, the committee has elected to follow common usage and use *alluvial fan* as the generic term for any of these categories.

In NFIP Regulations, CRF 44, §59.1, when the form and position of the flow paths is so radically uncertain that the risk of flooding at a place cannot be estimated through traditional procedures, a characteristic that frequently is associated with alluvial fans, this type of flooding is called *alluvial fan flooding*. Deviation from the traditional flood paradigm is further compounded on alluvial fans subject to debris flow hazard, that is, where the base flood is not caused by runoff but by a debris flow with triggering mechanisms, flow characteristics, and probability of occurrence that are completely different from those assumed in hydrological models of flood behavior.

Because the designation that an area is subject to alluvial fan flooding sets in motion specific, restrictive federal regulations, the determination by FEMA that an area is subject to alluvial fan flooding rather than ordinary flooding during the 100-year flood can be controversial. Thus, this report is in part an attempt to clarify what *alluvial fan flooding* means by providing a more precise definition and by describing how to apply the definition through the use of field indicators. The current chapter discusses this essential attribute of alluvial fan flooding in the section entitled "Origin of the Problem."

FEMA developed a procedure for estimating flood risk in environments subject to alluvial fan flooding (Dawdy, 1979; FEMA, 1990). In particular, the method predicts the extent of a fan-shaped area subject to a 1 percent chance of flooding in any year, as well as the average speed and depth of such a flood. Application of the *alluvial fan flooding* definition, the associated regulations, and the procedure for risk estimation together aroused considerable opposition from floodplain managers and other interested parties in some (but not all) communities that participate in the NFIP. This conflict led, eventually, to confusion and mutual suspicion.

Recognizing the need to resolve the conflict, FEMA requested the appointment of a National Research Council committee to study the issue of alluvial fan flooding. In particular, the committee was asked (1) to clarify, as necessary, the definition of alluvial fan flooding contained in section 59.1 of the NFIP regulations, (2) to specify criteria that can be used to determine whether an area is subject to alluvial fan flooding; and (3) to provide examples of applying the revised definition and criteria to real situations.

The Committee on Alluvial Fan Flooding met four times to study these issues and conduct a series of field visits during which it consulted with FEMA staff, floodplain management experts,

and local public officials (see Box 1-1). The committee's approach involved examining the hydrologic and geomorphologic processes that characterize flooding on alluvial fans in a range of varied environments. Understanding these processes from a natural science perspective
provides the committee's basis for evaluating how current NFIP practice might be improved to better characterize flood hazards and to more accurately delineate zones of flood risk. The committee's revised definition of alluvial fan flooding is presented in this chapter. Later chapters provide an overview of flood and sedimentation processes on alluvial fans (Chapter 2), describe field indicators and methods to delineate hazard boundaries based on the revised definition (Chapter 3), and give examples of applying the field indicators to specific sites (Chapter 4). Chapter 5 presents the committee's conclusions and recommendations.

Because the establishment of this committee was requested by FEMA, that agency and its consultants are the primary audience for this report. However, the committee hopes that communities participating in the NFIP, other agencies, and floodplain management professionals in general will also appreciate this effort to better manage natural hazards in alluvial fan environments.

ORIGIN OF THE PROBLEM

Following a series of damaging floods in the southwestern part of the United States during the 1970s, FEMA sought a new approach to flood risk assessment in areas where flow paths are difficult to predict. Pictures of these floods are shown in FEMA Document 165 (FEMA, 1989), and the images are memorable (see Figure 1-1): water and debris flows along new paths not anticipated by planners and residents of the normally dry landscape, automobiles crushed by boulders, a house full of sand. These pictures depict a type of natural hazard different from ordinary riverine flooding. The hazard on active alluvial fans is less foreseeable, more difficult to control or resist, and more dangerous.

Many of the pictures in FEMA Document 165 are of flooding on alluvial fans. The term applied by the NFIP to this image was *alluvial fan flooding,* and the purpose of Document 165 was to explain how such floods occur. As our understanding changes of how floods occur, regulators develop new policies that eventually become formalized by the writing of regulations. Special rules were thus promulgated to regulate development and to set insurance premiums in areas subject to alluvial fan flooding. Conflict arose when Flood Insurance Rate Maps (FIRMs) prepared by contractors to FEMA were criticized by some participating communities. Most criticism focused on the underlying assumption (or "default assumption") in the FEMA procedure that flooding on alluvial fans is completely unpredictable, an assumption that is not always appropriate (French et al., 1993).

The Problem of Delineating Flood Hazards on Alluvial Fans

Faced with an increased need to map flood risk on alluvial fan areas in the 1970s and 1980s, and needing a method that can be applied at reasonable expense, FEMA adopted an analytical technique proposed by Dawdy (1979). The procedure uses a general mathematical formula (known as the conditional or total probability equation) to describe the probability of an

---

BOX 1-1
SELECTING THE SITE VISIT LOCATIONS

To respond to FEMA's charge, the committee gave a great deal of consideration to the geographic locations of its site visits. The committee had support to meet four times during the study process and with those four meetings decided to focus on geographic areas where the implementation of the NFIP to alluvial fans has been either particularly challenging or has met with moderate success. Many sites were considered and with only four opportunities to meet the committee certainly could not visit every location of interest. The experience of the committee's members and extensive use of the published literature of course expanded our knowledge base considerably, but there is special value to site visits because they allow an opportunity to talk directly with the people involved.

The first meeting was held in Arizona where FEMA has encountered notable resistance to its regulatory approach for alluvial fan areas. The second meeting was held in southern California where there is a wider degree of acceptance of FEMA's approach, at least as far as the mapping methods are concerned. The third meeting was held in southern Utah where FEMA was endeavoring to identify hazard areas on debris flow fans, an important issue. The fourth meeting was devoted primarily to writing this report but the committee brought in guests to talk about the difficulties of identifying flood hazards in the arroyos near Albuquerque, New Mexico.

Had there been more time, the committee of course would have liked to talk with experts carrying on important work at a number of other locations. These include the immense flood control facilities being constructed in Clark County, Nevada; the technical research being conducted by the Department of Energy at the Nevada Test Site and other federal lands; the hydrologic studies being performed as part of the proposed Yucca Mountain nuclear waste repository; the arroyos of western Texas; the composite fans in the Death Valley National Monument; the alluvial fan flood hazards in the Navajo Reservation; the large scale alluvial fans along the Tahachapee Mountains in Kern County, California; urbanized fans such as in Wenatchee, Washington; and the effects of water-well induced ground subsidence on channel incision in the alluvial fans of central California.

---

event given a knowledge of some associated event. The procedure calculates the depth and velocity of the flood that has a 1 percent chance of occurrence at any point on a fan-shaped region, given a knowledge of the peak flow-frequency relationship at the apex of the fan. The procedure is based on only a few postulates and is attractive in the simplicity of both its conception and its implementation.

In its initial form (Dawdy, 1979), the procedure assumes:

- That the peak flow-frequency has been estimated for the apex of the fan.
- That the alluvial fan is shaped like a sector of a cone, all of which is subject to flooding, and that channels move across its surface at random during floods or from flood to flood (i.e.,

FIGURE 1-1 Alluvial fan flooding: High-velocity flows battered homes in Ocotillo Wells, California, during the September 1976 flood caused by Tropical Storm Kathleen (top). Fast-moving floodwaters caused scour, erosion, and structural damage to numerous Rancho Mirage, California, homes in September 1976 and July 1979 (middle). Large volumes of sediment can be deposited by floodwater during the course of an alluvial fan flood event (bottom). SOURCE: FEMA (1989).

that their behavior is completely uncertain). The procedure ignores possible incision or stability of channels on alluvial fans. At any distance down the fan, the channel has an equal probability, over the long term, of intersecting any part of a contour in a flood, and this probability is proportional to the ratio between the widths of the channel and the total width of the fan at that radial distance.

- That individual floods remain in single channels in which the flow occurs at critical depth and velocity (i.e., velocity is proportional to the square root of depth) and has a width-to-depth ratio of 200.

Although the original outline of the procedure was based on the assumption of complete uncertainty about the behavior of channels, the recommendations for application were quite flexible (Dawdy, 1979) and left open the importation of other concepts and data to constrain the generally applicable probability theory (e.g., Mifflin, 1990). FEMA adopted this method in an appendix to its *Guidelines and Specifications for Study Contractors*, calling it, in early versions, "Alluvial Fan Studies" (FEMA, 1985), and, in the latest version, "Studies of Alluvial Fan Flooding" (FEMA, 1995). The simplest form of this method was eventually codified into a computer program called FAN (FEMA, 1990), which likened alluvial fan flooding to rolling balls down a cone. In this form, the procedure can be followed by anyone, even with little or no knowledge of alluvial fans and their flooding characteristics. The FAN manual leads the practitioner through the procedure required to map zones of flood risk on the basis of only (1) a cursory identification that the site of interest lies on an alluvial fan, (2) measurement of the apical angle of the fan for computing its width at any radial distance, and (3) choice of a peak flow-frequency curve from regional data or a similar source.

When delineating flood hazard boundaries, the assumption of complete randomness in channel behavior may be relaxed if some field information is available that will allow the conditional probability equation to be solved with other constraints. But there does not appear to be any practical process, other than the review by consultants to FEMA, for deciding whether or which modifications should be applied in a particular circumstance.

Despite the elegance of its formulation, the FEMA procedure has been resisted in an important number of locations where it has been applied, particularly in those communities with the financial and technical resources to mount a challenge. The resistance arose for a number of reasons:

- Misapplication of the procedure to locations that were not alluvial fans or to fans or portions of fans that are not subject to flooding.
- Disparities between how floods occur on a particular fan and the assumptions of complete randomness.
- Assumption that the hazard is dominated by rainfall-runoff without recognition of the debris flow hazard in flood insurance studies.
- Too little investment being made in field identification of the conditions of flooding, leading to later huge expenditures for litigation and field surveys.
- The mismatch between the rather inclusive actuarial goals of the original procedure and other traditional uses of the resulting FIRM, such as floodplain management or hazard mitigation.

A number of professionals in the field have taken exception to the formulation of FEMA's risk delineation procedure (e.g., French et al., 1993; Zhao and Mays, 1993), asserting, for example, that alluvial fans do not necessarily behave randomly and that floods are more likely to follow previous flow paths; that flows do not occur at critical depth; that the width-to-depth ratio is different from the assumed value of 200; or that the method, as adopted, is unrealistically simple and therefore easy to misapply. Although specific applications of the method may be deserving of such criticism, the original formulation of the problem (Dawdy, 1979, equation 6) is quite general and sound for calculation of a conditional risk. This same mathematical approach has been used by the U.S. Army Corps of Engineers to analyze the flood risk of a levee failure while dealing directly with the uncertainties inherent in such an occurrence (USACE, 1992, 1994). If the conditional probability equation is to be applied to alluvial fans, however, there is need for a more flexible and realistic approach to the definition of flow path uncertainty, best obtained perhaps, by field evidence for the nature and spatial distribution of processes. A more realistic, process-based approach to flood hazard delineation on alluvial fans remains a challenge to the technical community. However, a mathematical framework for moving from process to the delineation of risk zones has already been correctly set forth by FEMA. Further discussion of methods to delineate flood hazards on alluvial fans, a secondary charge of this committee, is discussed in Chapter 3.

If it is determined that a particular flooding source does not match with the default assumptions of FEMA's alluvial fan method (for example, when the potential for channel movement is not random) the default risk delineation method is inapplicable. However, this does not mean that the area is necessarily free from alluvial fan flooding as discussed next.

The Problem of Defining Alluvial Fan Flooding

The following definition is published in the NFIP regulation (section 59.1):

Alluvial fan flooding means flooding occurring on the surface of an alluvial fan or similar landform which originates at the apex and is characterized by high-velocity flows; active processes of erosion, sediment transport, and deposition; and unpredictable flow paths.

The primary purpose of this definition was to identify the existence of and extend jurisdiction over flooding situations that may have been excluded or improperly dealt with under prior regulations. According to the NFIP rules, alluvial fan flooding is a type of flooding that is recognized by characteristics that distinguish it from ordinary flooding. Almost everyone that the committee heard from held the view that there is a strong correlation, or even an exclusive relationship, between the term *alluvial fan flooding* and flooding that occurs on an alluvial fan. By this reasoning, if one determines that an area is not an alluvial fan, then it is not subject to alluvial fan flooding. The current definition, however, states that alluvial fan flooding can be present not only on alluvial fans but also on the rather ambiguous category of "similar landform[s]." Such reasoning leads to conflicts. For example, people who believe that alluvial fan flooding is flooding that occurs only on alluvial fans and who obtain technical advice that their community contains no alluvial fans will therefore conclude that there are no areas subject to alluvial fan flooding in their community. FEMA, from its different perspective, might respond that within the community there are landforms similar to alluvial fans and that these experience alluvial fan flooding as explained in the regulatory definition. The community may counter that the similar landforms are inactive pediments and not alluvial fans; to which the response might be, " Just because you call something a different name, does not mean it is not subject to the severe hazards associated with alluvial fan flooding." The community's reply may then be that the amount of peril and the degree of uncertainty posed by flooding in its case are less than in other regions of the country that clearly have alluvial fans on which processes envisaged by FEMA are active and extensive. This hypothetical interchange typifies many of the discussions heard by the committee during the open, consultative portions of its meetings. There appeared to be a willful lack of reflection on the meaning of words, a general confusion about the diversity of flooding and sedimentation processes that occur on alluvial fans in a range of alluvial fan environments, and a lack of knowledge about information that could be quickly and inexpensively obtained through field examination of particular sites before hazard identification and delineation are carried out.

THE COMMITTEE'S RESPONSE

As a result of the committee's site inspections, the members' own field experience, consultations with many experienced individuals, including FEMA staff and its consultants, surveys of the literature, and extensive discussions at four meetings, the committee produced the following five products, which are presented in this report:

1. A revised definition of alluvial fan flooding, including criteria that can be applied by FEMA to identify those parts of alluvial fans that require special regulatory oversight to deal adequately with the uncertainty in flood processes. The committee also provides discussion of the constraints within which the alluvial fan flooding concept could be extended to other landforms. This definition is presented and analyzed in the latter part of Chapter 1. Although alluvial fan flooding is a general term that can involve flooding over an entire fan surface, the FEMA mandate is to determine the extent of flooding associated with a flood having a 100-year recurrence interval (i.e., a 1 percent probability in a given year). Hence, the term alluvial fan flooding is used in two ways. In the geomorphic sense, it can be any flood on an alluvial fan. But in the FEMA sense, it is the distribution of 100-year flood water on the fan. The reader is cautioned that the term is used in both ways, including in this report.

2. A description of the flood and sedimentation processes that build alluvial fans and contribute to the existence of alluvial fan flooding (Chapter 2). The chapter also illustrates why the recognition of how floods occur is significant for risk assessment. An understanding of process issues is necessary for realistic and flexible regulatory practice and flood risk assessment in a range of environmental conditions. For example, the process approach to understanding what is distinctive about alluvial fan flooding at a particular location and the conclusion that both the process and the channel form on alluvial fans respond to the environmental history of the site allow a quick and simple mapping of process zones on alluvial fans in the manner demonstrated in Figure 1-2. Such a map, which can be made quickly and inexpensively using methods described in Chapter 3, outlines areas that require various forms of attention from the point of view of flood

risk assessment and provides guidance about how to deal with uncertainty in methods such as those of Dawdy (1979) and Mifflin (1990) and the procedures discussed at the end of Chapter 3.

3. A description of indicators that are used to assess whether criteria indicative of alluvial fan flooding are present and which allow an investigator to discriminate between those zones of a fan where flow paths are uncertain and other zones where flow path uncertainty is unimaginable in the current range of environmental conditions (Chapter 3). This type of evidence suggests ways of dealing directly with uncertainty in flood risk assessment by (a) indicating realistically those areas that are truly subject to flooding, and (b) showing the nature of the flooding phenomenon that is being assessed. This chapter also illustrates how the alluvial fan flooding problem can be broken down into a set of questions for which there is already an established body of scientific material (e.g., channel bank stability) (USACE, 1992).

4. A demonstration of the use of the committee's definition of alluvial fan flooding for specific locations (Chapter 4), either field sites that were visited by committee members or sites described in detail in the scientific literature. The examples illustrate that various levels of effort yield answers of varying detail, but that even a one-day field examination can yield valuable insights for assessing flood hazard zones.

5. A framework that suggests the appropriate direction to advance our ability to delineate more accurately those parts of an alluvial fan that are subject to flooding by the 100-year flood by dealing directly with flood process uncertainty (Chapter 3).

THE NFIP DEFINITION OF ALLUVIAL FAN FLOODING

As noted earlier, the NFIP defines alluvial fan flooding as " flooding occurring on the surface of an alluvial fan or similar landform which originates at the apex and is characterized by high-velocity flows; active processes of erosion, sediment transport, and deposition; and unpredictable flow paths."

This definition emphasizes a type of flooding, not a landform, and thus is inherently difficult to translate into the regulatory setting. Defining the hazard more explicitly in process terms emphasizes that a variety of flooding processes with varying distributions and levels of intensity occur on alluvial fans; because of the range of environmental conditions in which such floods occur, a degree of flexibility is needed in defining and quantifying them. For emphasis and elaboration, the primary elements of the current NFIP definition are paraphrased here:

•   Alluvial fan flooding has an unpredictable flow path. The perceived channel, if there is one, may not be the actual conveyance route for water during a flood.

•   Alluvial fan flooding occurs on the surface of an alluvial fan or similar landform for which the spatial domain that is subject to flooding may extend over a larger area than the floodplain as determined by the traditional hydrologic paradigm.

•   Alluvial fan flooding has velocities high enough to erode new channels for floodwaters. Similarly, such erosion may undermine adjacent buildings and destroy them even though the water does not get deep enough to cause inundation.

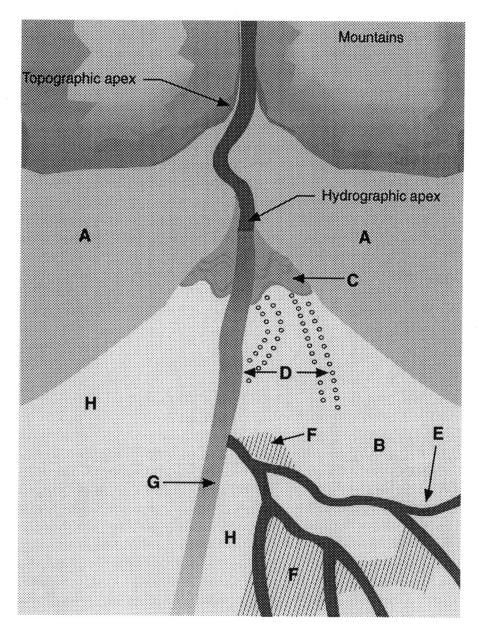

FIGURE 1-2 Example of a map that can be used to indicate areas requiring various forms of attention in flood risk assessment. The areas with solid shading are recognizable channels; the darker ones have stable forms and positions; the lighter shaded ones have the capacity to change form or position. A is an old fan surface that has been entrenched and does not receive runoff or debris flows from the mountain source area, and is not being undermined. B is a surface that is entrenched (but stands at an elevation below that of A), and will not be flooded or invaded by channels, which can become subject to these hazards if the current channel becomes blocked by a debris flow deposit. C and D are respectively bouldery lobes and levees indicating deposition by debris flows within and along channels. E denotes distributary channels that show no evidence of major scour, fill, migration, or avulsion during recent large floods and can convey all or most of a 1 percent flood, as indicated by reasonably applied flood conveyance equations. Areas indicated with F are subject to sheetflooding. G is a channel with signs of recent migration and for which future behavior is highly uncertain. H is a surface which is subject to overbank flooding, channel shifting, or invasion from a distributary channel that might erupt from G, and hence is the surface subject to alluvial fan flooding, as defined in this report. Further details of these processes and forms are given in Chapters 2 and 3.

• Alluvial fan flooding transports large volumes of sediment, the deposition of which may influence both the location and the direction of flowing water during a single flood event. This phenomenon is part of the reason for the "unpredictability" of the flow path.

The word "unpredictable" is troublesome in this context, and should be read only in a relative manner. There is always uncertainty associated with the prediction of how floods occur. Thus, in a sense, all flooding is "unpredictable." In the case of riverine flooding, which presumes a stable-bed condition, however, we can set aside this uncertainty because there are established procedures to predict how such floods occur. In the case of the alluvial fan to which the FAN computer program (FEMA, 1990) is applicable, we simplify the uncertainty by assuming that flow path behavior is random and a flood has no greater or less chance of following an established flow path than it has of cutting through the neighbor's backyard. The implication of the definition is that for alluvial fan flooding the flow path behavior is so indeterminate that we cannot set aside the uncertainty and still achieve a realistic assessment of the flood risk.

The existing regulatory definition does not describe an alluvial fan but rather a type of flooding that may also occur in nonalluvial fan areas (see Figure 1-3). For example, sediment movement may significantly affect flood flow behavior in river delta areas, and flow velocities high enough to cause erosion and deposition are common in alluvial river floodplains. After the 1993 Missouri River flood, sand deposits of 2 feet depth or greater covered 60,000 acres of adjacent farmland, causing damages in excess of more than $100 million (Interagency Floodplain Management Review Committee, 1994). On the other hand, there are alluvial fans that have well defined channels and may not be subject to alluvial fan flooding as defined above. The choice of the term is responsible, in large measure, for the confusion surrounding the definition issue. As it stands, the existing definition is vague and potentially very inclusive.

During the hazard identification process, the question is asked whether an area is subject to alluvial fan flooding. The answer affects both how flood zones are delineated (FEMA, 1995) and the rules that apply. Although the way Special Flood Hazard Areas (SFHAs) are delineated for alluvial fan flooding may differ from that for ordinary flooding, the regulatory importance of alluvial fan flooding is realized during mitigation. Section 65.13 of the NFIP regulations precludes the removal of a SFHA based solely on the elevation of a structure above the estimated flood stage or the placement of fill that creates such a condition. It also sets forth standards for structural flood control measures to remove the zone designation by mitigating the flood hazard.

Through discussion with representatives of FEMA and the review of various documents, the committee has identified several regulatory difficulties that originate from using the riverine flooding paradigm in alluvial fan flooding situations:

1. Issues related to Letters of Map Amendment (LOMAs). A LOMA is issued by FEMA when a parcel of land is inadvertently included in the SFHA because of the limitations of map scale or topographic data. By certifying that the finish floor of a structure is higher than the adjacent base flood elevation (BFE), the parcel can be removed from the SFHA. For areas subject to debris flows, extreme deposition, or shifting of flow channels, such an approach may not be appropriate, however, because the hazard still exists. Because LOMA applicants are allowed to contact FEMA directly, the agency has no way of determining whether a LOMA should be granted or not apart from the alluvial fan flooding designation.

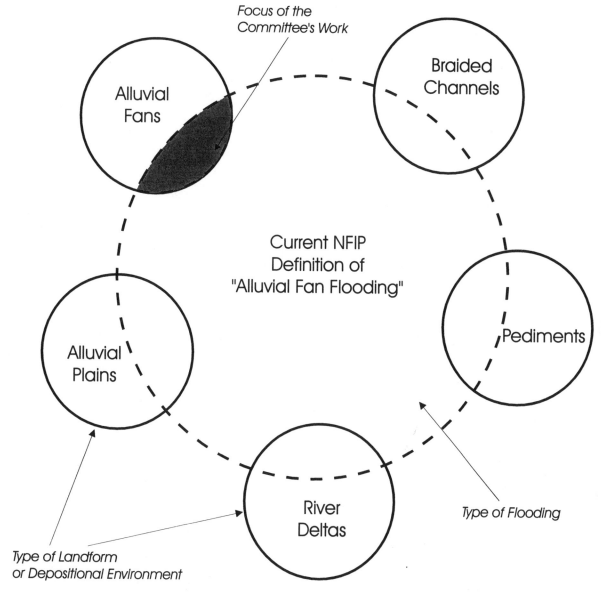

FIGURE 1-3 Committee's definition of alluvial fan flooding as it relates to various depositional environments.

2. Issues related to Letters of Map Revision (LOMRs) based on fill. A LOMR based on fill is issued when a parcel of land is raised by adding layers of soil so that the finish floor of a building is higher than the adjacent BFE. Since applicants for this type of LOMR can also contact FEMA directly, the same concerns exist as for a LOMA, that is, the processes of erosion and deposition and the impact force of debris, may still result in damage to the building.

3. The perceived flow path issue. Because of the convex cross-slope of many alluvial fans, floodwaters leaving the historical or perceived channel may follow a new direction and inundate areas distant from the channel. The alluvial fan flooding designation enables the inclusion of these

distant areas by consideration of the scenario where all or part of the flow leaves the main channel.

4. The sandbar issue. For a wide, shallow wash, a base flood elevation determined using backwater conveyance calculations such as those performed by computer program HEC-2 (Hydrologic Engineering Center, 1990) might indicate local areas within the wash that are higher than the computed water surface and could therefore be shown outside the SFHA. This, according to FEMA, could allow someone to build a house on a sandbar in the middle of a braided alluvial wash. The *alluvial fan flooding* designation discourages this by including the sandbar within the area subject to flooding and disallowing the removal of the designation until some type of reliable mitigation is implemented.

5. The split flow issue. Because of variations in channel cross section shape during an event or because of blockage by sediment, a network of channels that progressively splits into smaller channels may not reliably distribute the flow of water and sediment in such a way that all of the channels remain stable and protect the higher interchannel areas from flooding. For the case where the channel network fails, the *alluvial fan flooding* designation enables one to identify on the FIRM that the interchannel areas are at risk of being flooded even though they might be a meter or more higher than an adjacent channel.

Because these regulatory difficulties exemplify specific weaknesses in the riverine flooding paradigm, they portray some of the essential elements that make alluvial fan flooding a distinct type of flooding and they will help formulate the committee's revised definition.

IMPLICATIONS OF ALLUVIAL FAN FLOODING AS A DISTINCT TYPE OF FLOODING

Because of its character, alluvial fan flooding offers particular challenges to floodplain managers and regulators. Figure 1-4 compares the current paradigm that governs the analysis of ordinary riverine flooding, as viewed by the NFIP, to alluvial fan flooding. It contrasts the flood risk on two surfaces, labeled 1 and 2 in each case. The regulations require the identification of areas that have a 1 percent chance of being flooded in any given year. This process starts by the development of a graph showing peak flood discharge from the watershed plotted versus its recurrence interval. For most alluvial fan source areas, this step is tenuous at best and may be completely illusory in the cases where the events of significance are debris flows or where recent or frequent disturbance by fire makes it futile to view the watershed merely as a generator of independent rainfall-runoff events. However, this aspect of the uncertainty is not as controversial, perhaps because the methodology is widely used.

After discharge is specified, the riverine approach and alluvial fan flooding approach to analysis and risk prediction diverge. In the riverine case (Figure 1-4a), location of the flood within the spatial domain is assumed to be along the perceived or historical flow path, which is recognized as the main channel (Figure 1-4a(ii)). Much work is then necessary to identify the flood hazard by determining the relationship between depth and discharge (Figure 1-4 a(iii)). This is usually done using Manning's equation or a similar technique embedded within a step-backwater computer model. Finally, the flood hazard is delineated via the water surface elevation,

19

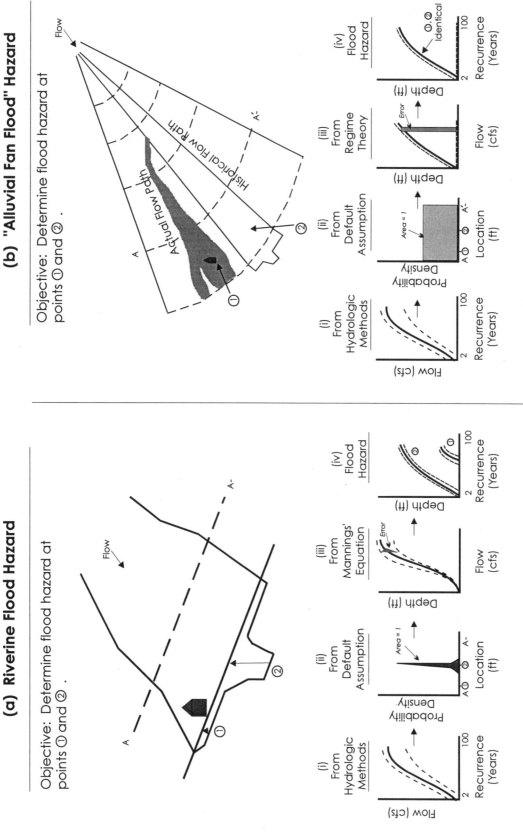

FIGURE 1-4 Comparison of the identification of flood hazards based on the traditional river floodplain paradigm with the identification of flood hazards considering alluvial fan flooding.

thus establishing a relationship between inundation depth above a point (such as the finished floor of a house) and the recurrence interval. Figure 1-4a(iv) shows the cases for the main channel and the higher surface (labeled 2 on the cross section), which is often called the *overbank area* or the *floodway fringe*. The stage recurrence graph developed in this manner always has some uncertainty, but it is usually not presented. The riverine approach provides a clear method that allows us to communicate about flooding and to make reproducible calculations of its severity. Over time it has become widely accepted and its weaknesses seldom questioned.

Figure 1-4b shows the analogous components of an analytical approach to alluvial fan flooding. First, the discharge recurrence relationship is estimated for the apex of the fan. Second, based on the knowledge that the perceived or historical flow path may not convey all of the water during a flood, it is assumed that the actual flow path has no greater chance of occupying the perceived channel than it does of straying to any location on the fan. This default assumption is shown in Figure 1-4b(ii). The relationship between depth and discharge is then determined using the method proposed by Dawdy (1979) (although a case can be made for altering this step by using the process-based knowledge described in Chapter 3 and an alternative solution to the conditional probability). Finally, the inundation depth and velocity are delineated based on the assumption that the entire fan surface is subject to flooding. The predicted degrees of flood hazard for the surfaces labeled 1 and 2 are identical because the procedure knows nothing of the differences between them. Real floods on alluvial fans are, of course, much more complex than this.

An important implication of this approach to the prediction of flood risk on alluvial fans is illustrated in Figure 1-5, which portrays a flood-prone surface with three distinct elevations, labeled 1, 2 (the main, recently occupied channel), and 3. In the traditional riverine flooding paradigm (Figure 1-5a) the historical channel is the main conveyor of the base flood and the computation of the water surface in "overbank" areas is based on the conveyance capacity of a single cross section that includes surface 2. This approach shows surface 3 as "wet" merely as a consequence of surface 2 being too small to convey the entire flood. Surface 1 is above the computed base flood elevation and could therefore be shown as outside of the 100-year floodplain on the FIRM. In such areas, we imply that the probability density function that describes flow path location within the lateral domain is narrow and strongly peaked, that is, that all of the flow behaves hydraulically as a single channel contained within a relatively narrow zone that does not shift during the event.

Figure 1-5b portrays the case of alluvial fan flooding where, during the base flood event, the channel might separate into two branches upstream of the cross-section, allowing a flow path to develop that invades the higher surface 1. After such a flow split occurs upstream, surface 1 may be flooded even during events smaller than the 100-year event depending on the specific behavior of a real sequence of floods. If the alluvial fan flooding paradigm is applied to the situation (Figure 1-5b), surface 1 is treated as a separate flow path, which it may well become during an actual flood. The corresponding probability density function for this situation shows three peaks indicating that each of surfaces 1, 2, and 3 have a finite chance of conveying water during a flood. Based on this potential, multiple scenarios are analyzed that represent the possible distribution of flows on the three surfaces. From this information, a separate baseflood elevation is computed for each surface via the conditional probability equation. The FIRM in this case would show that surface 1 is indeed subject to flooding. (Note: The probability density functions in these figures are shown to illustrate the difference between the two flooding perspectives. Showing the

(b) Application of "Alluvial Fan Flooding" Paradigm

◆ Examination of the site indicates that surface is ① subject to flooding.

◆ Using judgment, a probability density function is developed for the error in stage.

◆ All terraces are subject to flooding.

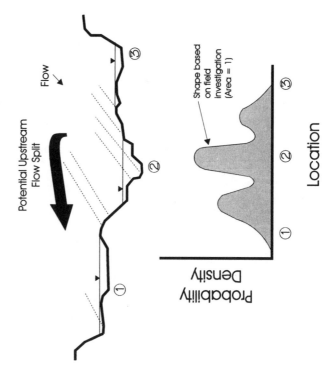

(a) Application of Traditional Flooding Paradigm

◆ The historical or perceived main channel is the primary conveyor of flow.

◆ The water surface evaluation in "overbank" areas is based upon the net conveyance of the total section.

◆ Surface ① is free from flooding because it is higher than the computed water surface.

◆ Surface ③ is inundated only as a consequence of the conveyance for surface ② being too small.

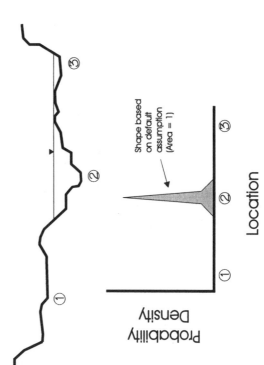

FIGURE 1-5 Application of flooding paradigms to a typical depositional environment.

general case, where a probability density function may be defined at each point of interest rather than for a cross section, would unnecessarily complicate the drawing of Figure 1-5.)

The alluvial fan flooding approach (Figure 1-5b) is a more realistic way of depicting the floodplain boundaries of even a river system to a large flood because the riverine flooding approach (Figure 1-5a) relies on maximizing conveyance in a preflood cross section to determine which surfaces are flooded and ignores the possibility that a surface can be flooded due to the redirection of flows at an upstream point. Thus, a key difference between alluvial fan flooding and riverine flooding is the implicit assumption of spatial consistency of the depth discharge relationship between adjacent surfaces. Based on physical processes, the alluvial fan flooding approach is the superior way for viewing how floods occur in general, and the riverine flooding approach is a special case where nothing out of the ordinary is going on. The question raised by Figure 1-5 is whether the presence of uncertainty in flood processes by itself constitutes alluvial fan flooding because failure to deal with channel changes by modeling scour, fill, and lateral movement (as the riverine flooding approach fails to do) results in grossly inaccurate delineation of flood hazard boundaries.

A previous National Research Council committee (NRC, 1983) concerned with the effects of process uncertainty in alluvial rivers was faced with the question of whether flood studies should use riverbed mobility models rather than fixed-bed models. Its conclusion was that the uncertainty introduced by ignoring the effects of sediment degradation/aggradation was no greater than the additional uncertainty introduced by the use of mathematical techniques. That committee's conclusion was that evaluating the effect of parameter uncertainty (i.e., the variation in channel roughness, geometry, and slope) was a suitable way to deal directly with process uncertainty in alluvial rivers. In other words, when executing the riverine flood paradigm it is often better to identify the potential impacts of error upon predicting the behavior of real floods than to strive for a more realistic approach in the hope that this error will go away. Simple approaches often yield satisfactory results. For example, the Flood Insurance Study for the City of Palmdale (FEMA, 1987, p. 8) recognizes this and contains the following counsel:

> Average depths of flooding were assigned based on standard hydraulic calculations through irregular cross sections. In many cases, the assigned average depth is not representative of the true degree of flood hazard. This situation occurs where the average depths are based on a wide cross section which encompasses one or more low flow drainage courses. The actual depth of flooding and, consequently, the true flood hazard will be greater adjacent to the drainage course.

The intensity of flood and sedimentation processes on many actively accumulating parts of alluvial fans is much greater and the frequency, magnitude, and suddenness of channel changes are more severe, however, than envisioned by the 1983 committee. In these more complex cases, the problem of channel location and flow distribution can be guided by the kind of process-based understanding, field evidence, and analysis techniques outlined in Chapters 2 and 3.

IMPLICATIONS OF ALLUVIAL FAN FLOODING
FOR FLOODPLAIN MANAGEMENT

The nature of alluvial fan flooding as currently defined by the NFIP has implications for both floodplain management and the mitigation of flood hazards. Consider the situation in Figure 1-5, which has been redrawn in Figure 1-6 to illustrate the implications of using fill to raise the level of surface 1. This surface is already above the computed base flood elevation, but it is clear that if one accounts for some degree of error, surface 1 could be completely inundated because it is protected only by a small berm in the natural topography. If ordinary riverine flooding is the appropriate paradigm to apply to the site (Figure 1-6a), filling surface 1 has no effect whatsoever on the computed bankfull flood elevation for surface 2 (the main channel). This would therefore be an acceptable strategy to reassure the concerned parties that the area is protected from flooding.

Viewed from the perspective of alluvial fan flooding (Figure 1-6b), however, where surfaces 1, 2, and 3 each have a chance of conveying all or part of the flood, protecting surface 1 in this manner would not be an allowable mitigation strategy. The reason is that eliminating surface 1 as a potential flow path, increases (theoretically) the frequency with which surfaces 2 and 3 get flooded. This is equivalent to increasing the base flood elevation for the n-year event and could therefore be an infraction of NFIP regulations (e.g., section 65.12). Similar actions that are generally considered to be good floodplain management practice would also come under question. For example, reinforcing a levee reduces the uncertainty about its failure potential during a flood and better protects areas behind it. But eliminating the flow path through a breach in the previously substandard levee could increase both the computed stage-frequency curve and the chance for other failures further downstream. These conclusions illustrate the disconnect between the alluvial fan flooding and the riverine flooding paradigms in the context of floodplain management.

For risk assessment under alluvial fan flooding, existing channels cannot be relied on to convey the 100-year peak flow, so their role is ignored. For riverine floodplain management, however, the channels are significant. They convey the smaller flood events, they indicate how floods have occurred in the past, and they define where future facilities may be located. The default assumption of a uniform risk (FEMA, 1995) or complete uncertainty across an alluvial fan is a formalized guess that allows one to delineate risk on the Flood Insurance Rate Map using a straightforward technique. A FIRM showing alluvial fan flooding hazards mapped in this manner is an expression of uncertainty or the absence of knowledge about floods, however, rather than an indication of how one might actually occur. Unlike a riverine FIRM, an alluvial fan flooding FIRM is of limited use for mitigation and management of flood hazards. By making a conservative trade-off in favor of what might happen, this type of FIRM ignores the importance of what has happened. If the uniform-risk FIRM is interpreted literally, then it can be argued using formal mathematics of the kind that underlies the existing FEMA procedure that any mitigation effort, short of complete channelization, increases the flood risk on another part of the fan. Thus floodplain managers are left with the peculiar responsibility of preserving uncertainty.

(b) Mitigation for Alluvial Fan Flooding

◆ Based on potential changes in the upstream flow path, surface ① has a conditional probability of being flooded.

◆ Mitigation of the flood hazard increases (theoretically) the frequency of flooding in surfaces ② and ③

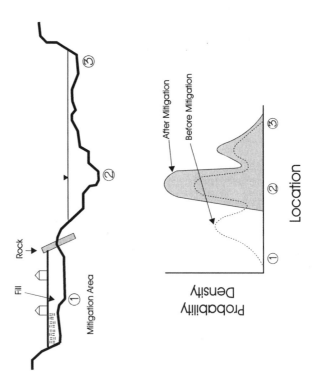

(a) Mitigation for Traditional Flooding Paradigm

◆ Considering the typical error of the traditional approach, surface ① is subject to flooding.

◆ Mitigation of the potential flood hazard does not impact the base flood elevations.

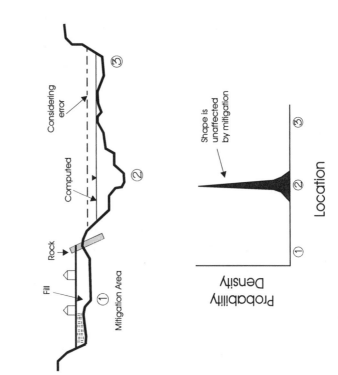

FIGURE 1-6 Mitigation from the perspective of two flooding paradigms.

THE COMMITTEE'S DEFINITION OF ALLUVIAL FAN FLOODING

This committee was asked to provide a new definition of alluvial fan flooding. Ideally, this definition could be used to divide flood hazard areas into two categories: those subject to alluvial fan flooding and those not. Of course, not all cases will fall clearly into one category or another. Although it is traditional to be conservative in the interest of safety, wholesale adoption of the alluvial fan flooding paradigm may not be a good thing because this approach brings its own set of weaknesses and regulatory traps. Furthermore, such blind caution could result in hundreds of cases where flooding sources that have been successfully mapped and managed as ordinary rivers will need to be completely reassessed for the sole purpose of consistently applying the definition. This would be unnecessary in many cases and could, ironically, undermine the original motivation for creating a new category of flood hazard for those cases that would otherwise be dealt with inadequately.

Because alluvial fans are the place where there is a strong historical connection to the breakdown of the riverine flooding paradigm, these landforms provide a regulatory partition that allows FEMA to concentrate on the most serious cases. This committee has therefore chosen to restrict the term *alluvial fan flooding* to apply only for alluvial fans. Floods with characteristics that fit the alluvial fan flooding concept but occur in nonalluvial fan environments are discussed as a separate, broader category of flooding.

The Committee on Alluvial Fan Flooding proposes the following definition, which considers the objectives of the original NFIP definition, the administrative concerns of the NFIP, and the criteria necessary to establish 100-year recurrence interval alluvial fan flooding as a distinct hazard:

> *Alluvial fan flooding is a type of flood hazard that occurs only on alluvial fans. It is characterized by flow path uncertainty so great that this uncertainty cannot be set aside in realistic assessments of flood risk or in the reliable mitigation of the hazard. An alluvial fan flooding hazard is indicated by three related criteria: (a) flow path uncertainty below the hydrographic apex, (b) abrupt deposition and ensuing erosion of sediment as a stream or debris flow loses its competence to carry material eroded from a steeper, upstream source area, and (c) an environment where the combination of sediment availability, slope, and topography creates an ultrahazardous condition for which elevation on fill will not reliably mitigate the risk (Figure 1-7).*

Alluvial fan flooding begins to occur at the hydrographic apex, which is the highest point where flow is last confined, and then spreads out as sheetflood, debris slurries, or in multiple channels along paths that are uncertain. The hydrographic apex may be at or downstream of the topographic apex and may change during a flood event due to deposition or erosion. Such flooding is characterized by sufficient energy to carry coarse sediment at shallow flow depths. The abrupt deposition of this sediment or debris strongly influences hydraulic conditions during the event and may allow higher flows to initiate new, distinct flow paths of uncertain direction. Also, erosion strongly influences hydraulic conditions when flood flows enlarge the area subject to flooding by undermining channel banks or eroding new paths across the unconsolidated sediments of the alluvial fan. Flow path uncertainty is aggravated by the absence of topographic

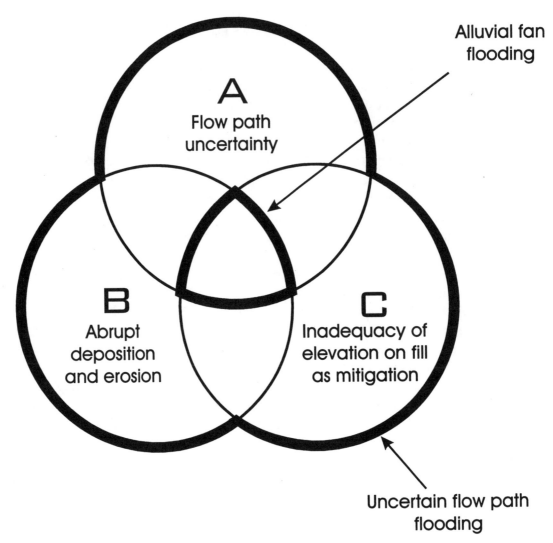

FIGURE 1-7 A, B, and C refer to the criteria in the committee's definition.

confinement and by the occurrence of erosion and deposition processes. Together, these characteristics create a flood hazard that can be reliably mitigated only by the use of major structural flood control measures or by complete avoidance of the affected area.

The potential for erosion and deposition, the related uncertainty in flow path behavior, and the imprudence of elevation on fill as a mitigation measure are joint and separate characteristics shared among many flood hazards on depositional environments other than alluvial fans, although not usually with the same intensity. It stands to reason that some of the same rules that apply to alluvial fan flooding should apply to this more inclusive type of flood hazard, termed *uncertain flow path flooding*. Flood hazards that meet only one or two of the criteria in the definition make up this third category.

To apply any definition in a regulatory context, the definition must be supported by criteria that can serve as standards, principles, or tests. These criteria, as reflected in indicators such as data, measurements, field evidence, and observations, are what floodplain managers and regulators use to apply the definition to a given situation. Erosion and deposition processes are essential criteria in judging alluvial fan flooding because they may affect hydraulic conditions. Where localized sediment deposition, bed form translation, and erosion produce changes in streambed elevation during an event that approaches typical depths of flow, the uncertainty introduced by these processes is significant. Consequently, floodwater surface elevations computed using preflood topography are not a comprehensive indicator of the true hazard for alluvial fan flooding situations as they are for riverine flooding.

Flow path uncertainty, which means that the perceived, historical channel or network of channels cannot be relied on to convey the base flood, affects the spatial domain subject to flooding through the creation of new flow paths and/or the subdivision of flows into multiple distinct paths as shown in Figure 1-5b. For a network of channels that exist prior to a flood event, uncertainty in the distribution of flows through the network must be considered in order to rule it out as a factor that might cause flow path uncertainty.

The criteria above frequently create a situation where the traditional formulas for mitigation, such as elevating a structure on fill, do not actually eliminate the hazardous condition. The unsuitability of fill as a hazard reduction strategy, however, is perhaps the most important characteristic distinguishing between riverine flooding and alluvial fan flooding.

In summary, the committee's revised definition limits alluvial fan flooding to flood hazard on alluvial fans. The committee recognizes that alluvial fan flooding is one type of flood hazard under the wider category of *uncertain flow path flooding*. Such hazards may have considerable uncertainty associated with their behavior and require means other than fill for reliable mitigation. Chapter 4 presents examples that illustrate how the definition applies to specific cases.

REFERENCES

American Geological Institute. 1987. Glossary of Geology. 3rd Ed. R. L. Bates and J. A. Jackson, eds. Alexandria, Va.: American Geological Institute.

Bedient, P. B., and W. C. Huber. 1992. Hydrology and Floodplain Analysis. 2nd Ed. Reading, Mass.: Addison Wesley.

Dawdy, D. R. 1979. Flood frequency estimates on alluvial fans. American Society of Civil Engineers Journal of Hydraulics Division, 105(HY11):1407-1413.

Federal Emergency Management Agency (FEMA). 1985. Appendix 4: Alluvial fan studies. Guidelines and Specifications for Study Contractors. Doc. no. 37. Washington, D.C.: FEMA.

Federal Emergency Management Agency (FEMA). 1987. Flood Insurance Study, City of Palmdale, California, Los Angeles County. Washington, D.C.: FEMA.

Federal Emergency Management Agency (FEMA). 1989. Alluvial Fans: Hazards and Management. Doc. no. 165. Washington, D.C.: FEMA.

Federal Emergency Management Agency (FEMA). 1990. FAN: An Alluvial Fan Flooding Computer Program, User's Manual and Program Disk. Washington, D.C.: FEMA.

Federal Emergency Management Agency (FEMA). 1995. Appendix 5: Studies of alluvial fan flooding. Guidelines and Specifications for Study Contractors. Doc. no. 37. Washington, D.C.: FEMA.

French, R. H., J. E. Fuller, and S. Waters. 1993. Alluvial fan: Proposed new process-oriented definitions for arid southwest. American Society of Civil Engineers Journal of Water Resources Planning and Management 119(5):588-598.

Hydrologic Engineering Center (HEC). 1976. Water Surface Profiles: Hydrologic Engineering Methods for Water Resources Development, vol. 6. Davis, Calif.: U.S. Army Corps of Engineers Water Resources Support Center.

Hydrologic Engineering Center (HEC). 1990. HEC-2 Water Surface Profiles, User's Manual. Davis, Calif.: U.S. Army Corps of Engineers Water Resources Support Center.

Interagency Floodplain Management Review Committee. 1994. Sharing the Challenge: Floodplain Management Into the 21st Century. Washington, D.C.

Mifflin, E. R. 1990. Entrenched channels and alluvial fan flooding. Pp. 28-33 in Proceedings of the American Society of Civil Engineers (ASCE) International Symposium of Hydraulics and Hydrology of Arid Lands. New York: ASCE.

National Research Council. 1983. Evaluation of Flood-Level Prediction Using Alluvial-River Models. Washington, D.C.: National Academy Press.

U.S. Army Corps of Engineers (USACE). 1992. Guidelines for Risk and Uncertainty Analysis in Water Resources Planning. Report. 92-R-1. Fort Belvoir, Va.: Water Resources Support Center.

U.S. Army Corps of Engineers (USACE). 1994. Risk Based Analysis for Evaluation of Hydrology/Hydraulics and Economics in Flood Damage Reduction Studies. Engineering Circ. EC-1105-2-205. Washington, D.C.: USACE.

Zhao, B., and L. W. Mays. 1993. Uncertainty analysis of the FEMA method for alluvial fans, Pp. 2098-2103 in Proceedings of the Annual Conference of the Hydraulics Division, Hydraulic Engineering, 1993. New York: American Society of Civil Engineers.

2

Flooding Processes and Environments on Alluvial Fans

FORMATION AND NATURE OF ALLUVIAL FANS

Alluvial fans develop where streams or debris flows emerge from steep reaches in which they are confined to relatively straight and narrow channels and flow into zones where sediment transport capacity decreases because of increases in channel width, reductions in channel gradient, or other influences. The channels on fans range from decimeters to several meters deep. These conditions develop at mountain fronts, in intermontane basins, and at valley junctions where there are major breaks in gradient or channel confinement, allowing both deposition of sediment and the lateral movement of channels to spread the sediment into a fan-shaped landform (Figure 2-1). Fan formation is particularly favored where sediment loads are high, for example, in arid and semiarid mountain environments, wet and mechanically weak mountains, and environments that are near glaciers or active volcanoes. Deposition is particularly rapid where there is a reduction in the transport capacity of a heavily loaded stream.

Alluvial fans occur in a wide range of environments, including the western and eastern mountains of the United States, western Canada, and various montane, arid, and volcanic regions around the world. In North America, most fans that have been subject to development are in the western mountainous regions. Fans occur in the Appalachian Mountains, but flooding on them has not yet been analyzed by FEMA because development pressure is not intense. However, minor local damage has occurred on some of these fans (Jacobson, 1993) and will no doubt increase as development pressure increases.

In the simplest cases of widely spaced stream or valley sources, fan geometry may be a sector of a simple cone emanating from a single, well-defined apex. In this simple case, a stream follows more-or-less a radial path down the cone, and the contours on the map of such a simple fan are convex downslope (Figure 2-1). Overall radial profiles are usually concave or virtually straight, and cross-fan profiles are convex. Where the sedimentary accumulations from several source areas encroach on one another, or where the deposition is forced by gradual widening or slope reduction along a valley, the simple conical fan shape may not be easy to identify (Figure 2-2). Coalescence may lead to a general accumulation of overlapping fans along a mountain front, called a *bajada* (Figure 2-3). At their downstream margins, fans merge with the smoother depositional topography of valley floors, river terraces, and lake and coastal deposits, and the

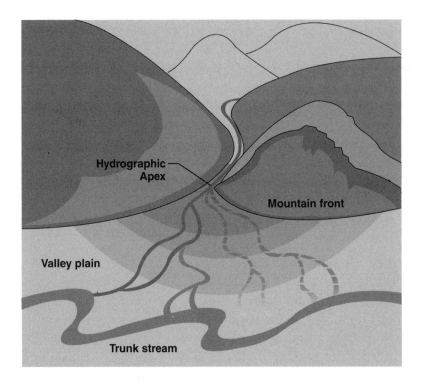

FIGURE 2-1 Sketch of a simple fan with single source and no incision. Contours are convex downslope and closer together near the apex. The dashed lines represent channels that have not recently been invaded by water or debris flows. The solid, sinuous lines emanating from the apex indicate channels that have conveyed flows recently.

channels may be small, shallow, and diffuse. Fans and bajadas are different from pediments, some of which are cone-shaped, in that a fan forms through deposition, whereas a pediment is a bedrock surface that is usually covered by a thin veneer of alluvium and colluvium.

Sediment may be transported to and across the fan by streamflow or debris flows. The latter are slurries with such high sediment-water ratios and concentrations of fine sediment that water cannot drain from them quickly enough to allow the sediment to settle out as traction load on the channel bed. Instead, the slurries travel at speeds of several to more than 10 meters per second (m/s) as dense viscous mixtures involving particle sizes from clay to boulders several meters in diameter.

Because the frequency, triggering mechanisms, size, and sedimentation processes of debris flows are so different from those of water floods, and the morphology and other clues about the nature of the flooding hazard on the respective types of fans are so radically different, it is necessary to distinguish between streamflow fans (Bull, 1977) and debris flow fans (Whipple and Dunne, 1992). Also, many fans are composites of stream and debris flow sediment. This chapter

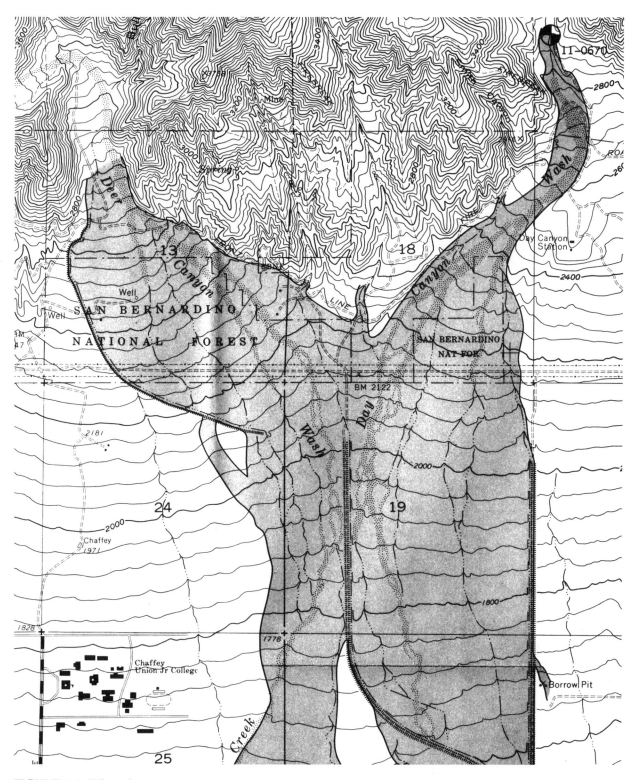

FIGURE 2-2 Where fans converge from multiple source valleys, the fan shape may not be obvious. The coalescing Day and Deer Canyon fans on a bajada along the southern slopes of the San Gabriel Mountains near Cucamonga, California, flooded in January 1969. SOURCE: Singer and Price (1971).

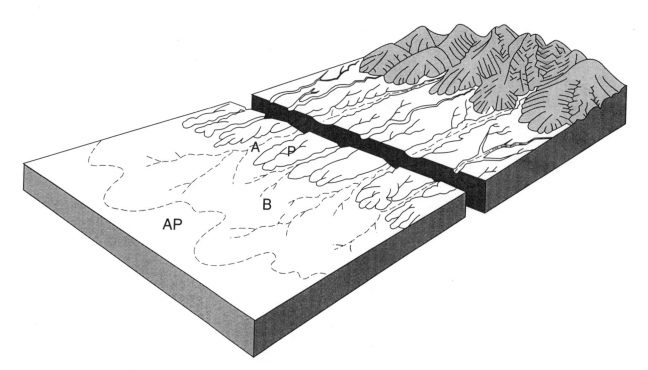

FIGURE 2-3 Stylized view of a bajada (B) showing alluvial fans (A) merging with an alluvial plain (AP). The bajada is formed by coalescing alluvial fans originating at gullies cut in a dissected piedmont (P) and by debouching on the fan piedmont. Such eroded fan-piedmont remnants commonly form the slopes above bajadas in the arid southwestern United States. SOURCE: Adapted with permission from Peterson (1981).

emphasizes the differences by using the terms *streamflow fan*, *debris flow fan*, and *composite fan*, unless referring to a generic accumulation of any origin as a *fan* or *alluvial fan*. This distinction is not widely made in the literature, where all fans are usually called *alluvial fans*, but it is important because recognition of the nature of flooding and sedimentation processes on a fan and an understanding of the difference in triggering mechanisms and therefore probabilities of debris flows and floods are crucial to the accurate interpretation and prediction of flood risk.

The continuum of fan types reflects the range of sediment transport and depositional processes that generate and modify the landform. Sediment supply and transport mechanisms on fans include debris flow, channelized water flow and sheetflood (extensive, shallow overbank flooding of water or mud). There is no sharp line differentiating channelized flow, and sheetflood. Some fans exhibit all three transport mechanisms, with the frequency and importance of each changing down the fan.

There is also a continuum in the intensity of the sedimentation processes and therefore in the activity of the fan-building and fan-modifying processes. Some fans are accumulating and changing rapidly under current climatic conditions; others are developing only slowly because changes in climate over recent millennia to centuries have caused their channels to deepen and stabilize. Still other fans have been subject to an intensification of flood and sedimentation hazard as a result of land use or engineering structures in the source area or on the fan itself.

Field evidence is an important source of information on the nature and intensity of the sedimentation processes that built the fan and is therefore critical for refining estimates of the nature, frequency, physical controls, and engineering significance of the flood risk on any fan. The morphology of the fan surface and the character of the deposits visible in the sides of channels indicate the relative contribution of water flows and debris flows to the flooding hazard on various parts of the fan. It can also be established whether changes in the governing geographical factors have changed the nature and distribution of the flood hazard during the history of the fan. It is important to realize that although fans vary in geometry and flooding characteristics because of the various combinations of their controlling factors, it is not always necessary to regress to a default assumption that there is no way to reduce uncertainty in the prediction of flooding processes on them. The copious field evidence available provides a means of reducing uncertainty about flood behavior, if it is properly interpreted.

Streamflow Fans

On fans that are actively forming from water-borne sediment alone, channels are usually braided, or multithreaded, from the apex. Deposition occurs on the channel bed in the form of bars on the margins or in the centers of channels. Rapid erosion of channel bed and banks is possible because of the loose, unconsolidated nature of the sediments. Thus, rapid erosion or deposition along a channel reach can alter the flow conveyance capacity during a single flood.

Bank erosion and lateral bar formation can force the channel to shift, while both bed aggradation and mid-channel bar formation can force water overbank and into new paths, so that channels divide and streams episodically abandon one or both channels. Channels may shift dominantly as a result of the accumulation of lateral bars, in which case they do not build up their bed or banks above the level of the surrounding surface. In addition to this gradual channel migration, sudden changes in flow path (avulsions) can occur due to overbank flooding. Even quite large and well-defined channels can be abandoned if a flood breaches one of the channel banks and water flows overbank in depressions between old bar deposits on the fan surface, often eroding a deep channel headward up to the source channel, which is then diverted. Particularly large, kilometer-scale changes in the positions of flow paths and active sedimentation zones can occur without the channel occupying or shifting across intermediate positions if the channelized and the overbank flow cause sediment to be deposited within and close to the channel, raising the bed and the channel margins above the surrounding fan surface (Figure 2-4). Breaching of the elevated banks in a large flood can allow the flow to travel toward the lower areas between channels or along the fan margins. Small shifts near the fan head can cause dramatic changes in channel position farther down the fan.

As one proceeds down the fan, the channels separate more frequently than they join, so that on average the channels diverge and diminish in width, depth, and discharge along a general flow thread during any one flood event. Despite the decrease in discharge, the reduction in width, depth, and gradient can force water overbank in many floods, and thin, unchannelized, relatively uniform expanses of water can cover large areas (*sheetflood*). Sufficiently far down some fans, most of the runoff occurs as a sheetflood, either generated locally on the fan or forced overbank by the diminishing conveyance capacity of the channels. The sheetflood itself is irregular with zones of concentrated flow giving way downslope to divergences and shallowing at which small

FIGURE 2-4 Oblique aerial photograph of an alluvial fan of the western slopes of the McDowell Mountains in central Arizona showing recent separation and braiding of channels downfan. The topographic apex is at the lower edge of the scene. Floodflow can inundate much of the area in the scene except for a few ridges of old-fan remnants. Courtesy of H. W. Hjalmarson.

fans of sediment are deposited. At the toe of most fans, sheetflood that has a relatively low sediment concentration because of deposition on vegetated surfaces can once again gain an erosive capacity if it is concentrated into a number of swales and small channels before entering trunk streams that drain the entire mountain front.

On streamflow fans where the sediment balance has turned negative, either at present or for some period in the recent past, the channels are deeper because sedimentation on their floors and margins is replaced by incision. The flow and sediment conveyance capacity increase because form roughness is less in the absence of aggressive bar growth. Thus, many of these channels are incised below the surrounding fan surface, and avulsion occurs less frequently or not at all in the current climatic and hydrologic regime. The separation of the flow into diverging, smaller channels is reversed, and one or a few trunk streams convey the floodflow to the toe of the fan. Because these major conduits are incised they are not so frequently diverted by mid-channel bar deposits and they do not shift across the fan as quickly as those on actively accumulating fans. Instead, they tend to gather local runoff generated on the fan surface because the rills and small channels produced by such runoff repeatedly erode toward the stable trunk channel. The channel network is slightly convergent downfan, and mapped contours show upfan re-entrants that reflect the incision. Some channel banks may be colonized by trees and bushes, adding another stabilizing influence. Because their surfaces are no longer accumulating sediment, such fans or parts of fans are said to be *inactive*.

Incised streamflow fans are particularly well developed in regions where a major climatic change has altered the conditions that favored sediment accumulation (e.g., the transition from glaciation to interglacial period in parts of the Sierra Nevada and the Cascade Range of Washington state, or from wetter to drier periods in the mountains of Arizona). They are also well developed in areas where fans have been steepened tectonically as in parts of southern California. In these cases, there is a strong isolation of deeply incised channels from the surrounding "fossil" fan surface. Thus the problem of recognition is complicated because all degrees of isolation occur, ranging from aggressive accumulation to deep incision. Chapter 3 describes field methods for identifying and mapping degrees of activity and for dating the time of latest activity on various parts of fans.

Intermediate cases of channel stability and confinement are particularly widespread and important to recognize and evaluate. They occur, for example, in diverging channel systems (*distributaries*) where the sediment balance of a reach is near-steady state. Such channels may gradually become shallower downfan until their floodwaters simply disperse as sheetflood, repeatedly spreading thin layers of sediment and water and building an apron of relatively well watered and fine sediment that supports thick vegetation. In some years, there is accumulation of sediment, and in others there is net removal, so the bed may rise or fall by a few decimeters, but neither the scour nor the filling trend persists for long enough to raise or lower the channel bed significantly in relation to the fan surface.

Alternatively, there may be a persistent but very gradual trend that causes the channel to rise, lower, or shift laterally at a rate that is difficult to detect with commonly available information (e.g., sequences of maps or aerial photographs, anecdotal reports, dating of recent sediments with buried artifacts). In other cases, a reach that has stabilized may be perturbed by runoff or sediment that enters it from an unstable reach upstream. Thus analysis of the stability of a reach requires taking a broad view of the potential for change in channels upfan. Spatial context is important in any analysis of flooding and sedimentation hazards on a streamflow fan. Hjalmarson (1994) provides an illustrated account of various distributary flow channels with a range of flow path stability and intensity of flood hazard.

Debris Flow Fans

Debris flow fans occur where strongly episodic sediment transport is triggered by collapse of an accumulation of weathered rock, soil, or sediment in a steep source region or by concentration of flow onto a steep accumulation of sediment that is then trenched rapidly in such a way that a high sediment concentration is developed with a mixture of sizes, including a significant proportion of fine sediment. The sediment-water ratio of the mixture must be so high that the flowing debris has a low permeability and water cannot drain out (upward) quickly enough to allow the water to separate from the sediment and the sediment to settle onto the bed.

The resulting poorly sorted slurry is dense and highly viscous and travels as a laminar flow except where agitated by waterfalls and cascades, by larger rocks in the bed, or by engineering structures. Observers often describe such flows as looking like wet concrete. Flows with intermediate sediment-water ratios and characteristics between those of debris flows and turbid water flows are sometimes referred to as *hyperconcentrated flows*.

Debris flows consist of the full range of sediment sizes supplied from the source area, and flows generated from rocks of different types within the source basin may contain different proportions of clay. The greater the proportion of fines, the greater is the internal strength of the flow because of "cohesive" bonding caused by electrical charges shared between clays and water films. Some flows are sufficiently dense and viscous to transport boulders; others leave the largest boulders behind. As the sediment-water ratio decreases (i.e., in more dilute flows), progressively smaller boulders settle to the bed and are deposited or transported as traction load in a turbulent flow.

The flow properties of the slurries determine the fate of the debris flows when they emerge onto the fan, the nature of sediment deposition, and the resulting morphology of the deposit. These properties depend on the magnitude of the discharge and the rheological properties of the debris, which in turn are controlled by its sediment-water ratio and clay content. Discharge rate, clay content, and sediment-water ratio of each debris flow are set by the generating mechanism and the particular combination of circumstances that trigger the flow. For example, a large rainstorm or snowmelt may generate landslides that fall into stream channels containing significant discharge, and the resulting mixture may produce a dilute debris flow. Collapse of wet debris into a steep channel network that already contains a large volume of fallen debris from centuries of slow mass wasting on adjacent hillslopes may result in scour of that accumulation into a particularly dense and viscous, boulder-charged debris flow. The volumes and peak flow rates of debris flows depend on (1) the magnitude of the water supplied from a rainstorm, snowmelt, lake outburst, or volcanic eruption, and (2) the volume of loose debris that is available to be liquefied by this water during the initial collapse, undermining and assimilation, or scour from the valley floor along the steep portion of the debris flow track. Thus, the debris flows that supply and mold any one fan have a probability distribution of discharges and rheological properties, which determine the nature and magnitude of flood risk. Fortunately, these aspects of flood risk can be read from the morphology of the fan and its source basin.

The range of rheological properties among debris flows emanating from the source valley usually accounts for differences in morphology on different parts of a single debris flow fan. Flows with the highest sediment-water ratios and therefore the greatest strength come to rest on relatively steep gradients (typically 6 to 8 degrees) on the upper parts of the fan in the form of bouldery snouts and levees. These deposits block channels scoured by water floods between debris flow episodes and divert later flows of water or debris into new channels. The result is a topographically rough surface of berms, lobes, and bouldery channel blockages on the upper parts of debris flow fans (Figure 2-5).

Somewhat more dilute and weaker flows travel through the steepest channel reaches, but deposit bouldery levees as their margins are slowed. If the peak discharge rate of a debris flow exceeds the conveyance capacity of the channel, its upper part is partially decanted overbank and it travels some distance across the fan surface until it becomes slow enough and thin enough to stop as a bouldery or gravelly sheet with a sharp edge. Stranding of boulders in levees and overbank sheets causes a progressive downfan reduction in the boulder content of flow deposits. The most dilute and weakest debris flows remain channelized as far as the lower parts of the fans, where gradients may be as low as 2 to 3 degrees. Some of these flows halt within the channel, raising its bed and lowering its depth, while others spread over the banks onto the surface of the

FIGURE 2-5 The upper part of this debris-flow fan in Owens Valley, California, shows a classic rough, bouldery surface. Courtesy of T. Dunne.

fan as the declining gradient reduces the conveyance capacity of the channel, forcing flow overbank. The result is a smooth surface with only an occasional boulder on the lower parts of a debris flow fan. On debris flow fans, streams are often confined to nondiverging, boulder-lined channels left by the debris flows, and therefore they neither shift across the fan nor overtop the banks in most cases, except on the lower parts of the fan where shallow channels were originally formed by the dilute, low-viscosity flows described above. Of course, if the debris lining the channels is gravelly rather than bouldery, the capacity for channel shifting and eventual realignment by water floods is greater.

Many channels on debris flow fans are single-thread depressions blocked at their upper ends by bouldery accumulations, so they are never invaded by stream floods or debris flows. Like alluvial fans, debris flow fans are subject to varying amounts of deposition and parts or even much of the fan may be inactive under the present climate. For example, the debris flow fans emanating from the east side of the Sierra Nevada in the northwestern part of Owens Valley have more or less ceased to accumulate since the end of the last glaciation in the mountains, and the oldest parts of the fans date from previous glaciations. Parts of fans debouching from the unglaciated southern Sierra and from the White Mountains on the western side of Owens Valley continue to receive debris flows in the modern climate. Descriptions of large debris flow fans in Owens Valley, California, are provided by Whipple and Dunne (1992), and smaller debris flow fans in a wetter environment are described by Kellerhals and Church (1990).

One approach to flood risk on debris flow fans concludes that even on active fans the probability of a debris flow is less than 1 percent in any one year, and therefore the "100-year flood" is not a debris flow but a runoff event. This is a generalization that fails to appreciate an important aspect of debris flow initiation, namely, that it is not an independent, random event in the same way that runoff floods are assumed to be. Debris accumulates in source localities and along stream channels over timescales from decades to centuries between failures that evacuate the debris (Benda and Dunne, 1987; Dunne, 1991; Reneau and Dietrich, 1991). Thus, a frequency count of dated debris flows in a region might indicate that the average frequency of occurrence is, say, 200 years per fan (with a probability of occurrence in any one year of 0.5 percent). However, if a geologist were to walk up any one of the source basins, he or she might find many potential failure sites and the channels below them to be occupied by thick layers of sediment that have accumulated since the previous debris flow occurred centuries earlier. In a neighboring valley, recent debris flows may have stripped such sediment from the valley and reset the clock so that the probability of debris flow is virtually zero for the foreseeable future. Thus flood risk estimates can be refined by first recognizing from field evidence that debris flows are the dominant sediment transporting agent on a particular fan and then examining the source basin to determine whether debris availability favors an enhanced risk of a debris flow in the event of a large rainstorm or snowmelt.

Composite Fans

Many fans are fed by both water floods and debris flows. Others were formed by debris flows under a different climatic regime and are now the sites of stream sedimentation and flooding only. Thus, both streamflow sediments and debris flow sediments and their associated morphologies attest to the nature of the flood risk on different parts of the same fan. The debris flow sediments are usually concentrated on the upper, steeper parts of the fans, producing a surface laced with berms, lobes, and channel plugs. The lower, streamflow part of the fan has the characteristics of an alluvial fan described above, although there may also be a contribution of dilute debris flow deposition on these distal areas. An indication of the relative contributions of debris flows and water floods can be obtained through systematic identification and mapping of the distribution of the two types of sediments on the fan surface and in vertical sections along the sides of channels.

Incised Channels on Fans

At the heads of some alluvial fans, channels are strongly incised in a fan-head trench, from which they emerge at some distance downfan to take on the character of a diverging braided channel network or a linear, boulder-leveed channel characteristic of debris flow fans, as described above. This report calls this point the *hydrographic apex*. Several reasons for the transition are identifiable in the field. The simplest case arises on a composite fan where episodic debris flow deposition produces a gradient that is steeper than that required by intervening floods to transport the sediment load supplied by the source basin or produces sufficient channel bank strength to confine water flows to depths sufficient to transport the sediment supply. In this case, the water floods will scour away some of the debris flow sediment, establishing a lower-gradient channel incised within the debris flow deposits. At some distance down the fan, where the gradient of the debris flow sediment surface has diminished, the required stream gradient intersects the fan surface and a single-thread or braided channel or a swath of sheet flooding emerges from the fan-head trench at what Hooke (1967) called an *intersection point*, that is, a transition between flow and sediment transport process regimes.

In other cases, trenching at the fan head or even over the entire fan may occur as a result of channel incision of older fan deposits, either because the sediment supply has diminished or because the transport capacity has increased owing to climate or vegetation changes within the source basin, or to tectonism. The roles of climatic change and tectonism in trenching the heads of fans are reviewed thoroughly by Bull (1991).

FLOODING PROCESSES ON ALLUVIAL FANS

Flooding on Streamflow Fans

Since streamflow alluvial fans typically occur in arid and mountainous environments, one of the first difficulties encountered in the quantification of alluvial fan flooding processes is the magnitude-frequency relationship for flows supplied to the apex. The sparseness of hydrologic monitoring stations in such regions and the shortness of most records render most estimates of probable flood discharges highly uncertain. Fans receive high water discharges from hurricanes or typhoons on the subtropical eastern sides of continents, and from more localized rainstorms or from intense, persistent snowmelt in mountainous western North America. In southern Europe, particularly in southeastern Spain, the most destructive discharges are again generated by rainstorms. In each of these regions, the history of hydrologic analysis and prediction has been one of surprises.

Stream flooding on alluvial fans differs from most riverine flooding in that the hazard not only derives from the inundation itself, but also is intimately connected with sedimentation processes. These latter have immediate impact during the flood itself, and they have long-term geomorphic influence through the rearrangement of sediment on the fan. High flood stages in channels are accompanied by high flow velocities, and by heavy loads of floating wood, and other debris in some environments. High velocities are promoted by the relatively steep, hydraulically simple nature of the channels. The flood hazard is markedly increased, however, by the potential for channel change during the flood itself. The loose bed material may be scoured several meters deep. On some fans the loose, unconsolidated nature of the sediments allows rapid channel widening by bank collapse if the flood persists for several hours or days. On others, the deposition of bars along a channel margin causes the channel to shift against the opposite, concave bank at rates of up to tens of meters per flood. Thus, rapid scour and filling of the channel cause changes in the channel conveyance capacity between and during floods.

The largest and most widespread threat arises, however, through the process of avulsion ("tearing away") in which water escapes from a channel by scouring a new path through the bank.

This process may begin by sudden bank collapse or by gradual overflow of water from the rising flood. As the relatively dilute surface water flows overbank, it often travels down a gradient that is steeper than the channel (because of the convexity of the fan cross section and the perched nature of some channels above the general fan surface) and so is able to pick up sediment and scour a new path. The water may also take advantage of a former channel, or a series of abandoned channel segments. In the short term, this process may be easily predicted if there are obvious low sections of bank or narrow levees separating the channel from much lower parts of the fan. However, it is difficult to anticipate all such weak sections of the banks and to predict the exact flow path that the diversion is likely to follow across the irregular fan surface, especially since the diverted flow has the transport capacity to modify that surface. On the time scale of decades, it is virtually impossible, with either field inspection or mathematical modeling of sediment transport, to anticipate the locations of in-channel deposition and bank erosion that might provoke avulsion. The problem is aggravated by the fact that a diversion on the upper part of the fan may alter flow paths on the lower part of the fan in ways that are independent of local fan morphology. Topographic changes in the channel network far upstream of a channel reach have the greatest potential for radically altering the risk of inundation, overtopping channel banks, or undermining a site downfan, but are probably the most difficult threat to anticipate and quantify.

Sheetfloods spread extensively on low-relief lower parts of fan, and as they decelerate they often deposit sheets and low bars of sand or gravel. Even though velocities and depths are low, inundation by turbid water can be very destructive.

Flooding on Debris Flow Fans

Debris flows are dense (approximately 1.8 to 2.0 times the density of water), viscous (approximately 10,000 times the viscosity of water), and fast (3 to 10 m/s (9 to 30 feet/s)). They can transport boulders up to several meters in diameter, either as individuals supported in the matrix of the flow or as dams of boulders pushed along at the front. Some debris flows consist of waves of slurry behind bouldery dams (Sharp and Nobles, 1953; Suwa and Okuda, 1983). Large woody debris, engineering artifacts, and vehicles are also transported by debris flows and can because blockage, flow diversion, and extra damage to houses and other structures downfan. As they approach their final deposition point, debris flow sediments acquire a finite yield strength that prevents them from draining away like water. They remain as permanent covers on fan surfaces, and are expensive to remove from urban areas or channels. (On the other hand, however, they are less likely than water flows to undermine and destroy a road, so once cleared the road is generally still useable.) In extreme cases, such as after particularly large debris flows on fans near active volcanoes, deposits may be so thick and extensive that they permanently bury settlements. The deposits also block drainage in valley floors and at tributary junctions.

Some aspects of the prediction of debris flow frequency and magnitude at the fan apex are more difficult than is the case for water floods, but other characteristics of debris flow occurrence simplify the problem. Within the United States, there are no monitoring stations with records long enough to provide a representative sample of debris flow occurrence on which a probability analysis might be based. A procedure commonly used by flood control agencies involves using records of runoff for prediction of a water flood peak with a 1 percent probability of occurrence, and then increasing the volume of that predicted flood peak to account for the observed sediment-water ratio in a recent debris flow (which usually triggered the concern by surprising the agency in the first place) (Brunner, 1992). Although such a procedure might give a reasonable answer for those hyperconcentrated flows generated by runoff processes, it is wrong to mix in a probability analysis the results of runoff processes (gauging station records of floods) with debris flows, which are the usually triggered by some form of mass failure.

A particularly misleading situation arises when the assumption of interannual independence that has been found to be a useful approximation for rain-generated and snowmelt floods is applied to debris flow occurrence. This is because the occurrence of a debris flow in one year substantially reduces the probability of future debris flows by removing the sedimentary accumulations required for their generation and growth (Benda and Dunne, 1987; Keaton, 1988; Keaton et al., 1988). Fortunately, it is often possible to identify through field observations those conditions that favor the generation and growth of debris flows. For example, deep accumulations of colluvium on bedrock indicate a relatively high probability of debris flow occurrence in comparison to that in a basin in which most of the colluvium was evacuated in a relatively recent meteorologic event, after a forest fire, or after a climatic change. Thick accumulations of sediment along channels upstream of a fan indicate that no debris flow has passed for a considerable amount of time and therefore that the conditions are evolving toward a failure that could convey large quantities of sediment from canyon floors to the fan. Such observations combined with a probability analysis of rainfall or snowmelt required to trigger a mass failure are required for estimating the debris flow risk at the apex. Estimating the probable magnitude is more time-consuming, since it requires documenting volumes of sediment in old debris flow deposits or in the valleys above the fan.

Avulsions of debris flows occur on boulder-rich fans and are particularly difficult to forecast because of the uncertainty about the magnitude and rheology of the next debris flow. However, some clues to the likelihood of an avulsion occurring can be obtained from field inspection of the morphology and sedimentology of the fan itself. In particular, useful indications of the avulsion potential might be provided by (1) the volumes of sediment susceptible to liquefaction in the source area (Keaton, 1988; Keaton et al., 1988) and therefore the likelihood of a peak discharge great enough to overtax the conveyance capacity of the channel for such a debris flow; (2) previous blockage of the main channel by bouldery berms; (3) relatively low channel banks near the apex or in the vicinity of any blockage in the main channel; and (4) the topography of the fan surface at these locations. Calculations of the channel conveyance capacity for debris flows with a range of rheology can be made for various channel cross sections down the fan to judge the potential for overbank flow and spreading (Whipple, 1992; Whipple and Dunne, 1992). At the distal margins of debris flow fans, low-strength flows often spread widely in a manner similar to sheetflooding on streamflow alluvial fans (Figure 2-6).

A particularly hazardous situation arises on debris flow fans around active volcanoes because of the huge volumes of sediment that can be liquefied and the persistence of the liquefaction. For example, the October 1994 lahar (volcanic debris flow) generated by a typhoon from the slopes of Mt. Pinatubo in the Philippines deposited approximately 50 million cubic meters (1.8 billion cubic feet) on an urbanized and cultivated alluvial fan. The North Fork Toutle River lahar generated by the Mt. St. Helens eruption of May 18, 1980, deposited approximately 100 million cubic meters (3.5 billion cubic feet) of debris (Fairchild, 1985). Such volumes

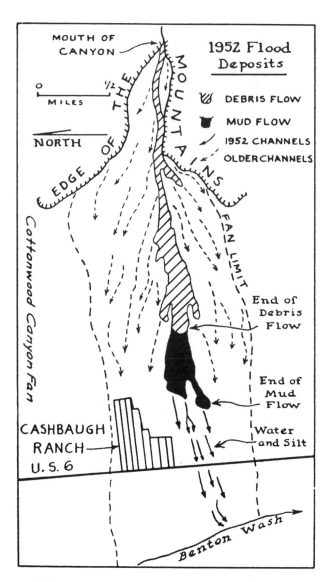

FIGURE 2-6 A 1952 debris flow flood on Cottonwood Canyon Fan, California. SOURCE: Reprinted with permission from Beaty (1963).

completely overwhelm the pre-existing channel system and topography of fans and create new pathways that are impossible to predict in detail.

Flooding on Composite Fans

Composite fans are subject to both streamflow and debris flow flooding. The relative importance of each varies between fans, between positions on each fan, between individual meteorologic events, and between episodes as climate fluctuates or as debris accumulates slowly for centuries after a period of debris flow activity.

In general, debris flow activity is most frequent on the upper parts of a composite fan, but a large, dilute, and therefore weak debris flow may remain channelized and mobile for tens of kilometers, and therefore traverse the zone that is dominated by streamflow activity. However, through geologic mapping of distinct debris flow and alluvial deposits it is usually possible to define and map the probable nature of the various types of deposition (bouldery, lobate debris flow deposits; channelized debris flow deposits with bouldery levees; tabular sheets of debris flow deposits; bars of streamflow gravel; and sheets of gravely or sandy streamflow deposits) and thus the maximum extent of debris flows on a composite fan. After examining the debris flow generation potential of the source area, it is also possible to make an approximate calculation of the maximum conceivable transport of debris flows downfan. It is also possible through geologic mapping and dating of deposits and examination of source areas to determine whether the debris flow activity that may have built most of a fan is still active or will generate much smaller debris flows than during the period of intense fan building.

An important issue for the prediction of flood risk on fans affected by debris flow is that the rearrangement of the channel system by a debris flow can cause a long-term change in the flooding hazard during stream flooding. Furthermore, after a debris flow has deposited large volumes of sediment on the upper parts of a fan, streamflow in the same or later events will spread the sediment far downfan, causing channel instability and sedimentation that reduce the flood conveyance capacity of channels.

CHANGE OVER TIME

Fans form over thousands to millions of years, during which time environmental conditions affecting their formation change more or less continuously. For this reason, they are always evolving, although parts (reaches) of them may temporarily attain a steady state in which their channel gradients can pass the supplied sediment and no further deposition occurs in that reach. Even in that state, the channels can shift across the fan and remold its surface by erosion and deposition.

However, part or the whole of the fan may become a zone of greatly diminished accumulation or in the extreme case may become a zone of net erosion with the entire channel system entrenched. The fan surface may thus be weathering and eroding, or it may have attained a steady state after a period of incision. However, some runoff and channel patterns develop on the fan after it ceases to be a zone of net accumulation.

Changes in the sediment balance and its spatial distribution can occur because of (1) externally imposed changes (e.g., of climate or land cover influencing sediment supply and transport capacity or of tectonism influencing channel slope and therefore transport capacity); (2) long-term evolution of the fan shape (especially its gradient) as it evolves over time in response to the sediment accumulation; and (3) shorter-term changes in form due to sediment accumulation (e.g., channel shifting accompanying bar accumulation, channel avulsion, raising of channel above the surrounding surface, deposition and channel filling by a single debris flow, runs of wet years caused by weather fluctuations, which may cause accelerated channel widening or shifting or reactivation of sediment sources producing debris flows that have been dormant for a long time).

Even if the fan is still a zone of active accumulation, the locus of that deposition may vary between fans or on the same fan during its lifetime. Distributaries develop most frequently where aggradation is intense and thus may shift the centers of active sedimentation across or along the fan. Deposition also may shift down the fan if the uplift rate in the mountain source exceeds the deposition rate at the fan head. This can cause trenching of the source canyon and the upper fan or intense episodes of debris flow and stream transport associated with increased runoff. In the latter cases, the channel may be entrenched below the upstream part of the fan surface and all deposition and attendant channel shifting or change occurs on the lower portions of the fan. Appreciation of such changes is important for understanding flood hazards on fans because (1) some of the changes occur sufficiently quickly to affect channels on timescales relevant to engineering analyses, so that it is important to know whether a particular fan is undergoing a change at the present time, and (2) geologically recent, but no-longer active, processes may leave a morphologic signal on the fan or in the source basin that must be interpreted in order to refine estimates of recent changes in flood hazard.

Fans grow in a variety of ways and the thickness generally increases at the slope transition formed at the fan toe as it is progressively buried (Hooke and Dorn, 1992). Deposits near the apex commonly are remobilized and redeposited farther down the fan on and below the fan toe. French (1995) gives a method of estimating the depth of sediment deposition at these slope transitions on fans. The most permanent deposition typically begins at the toe and propagates both up the fan and below the toe where the slope typically diminishes.

The evolution of the fan surface causes a difficult problem for the interpretation of field evidence concerning alluvial fan flooding and for the prediction of future flood risk. For example, if a part of a fan surface has not been disturbed by flooding or erosion for 15,000 years, its surface will have become weathered and covered by a soil-profile and vegetation (described in Chapter 3). The surface of such a fan will be very different from the surface of a nearby channelized and actively evolving area. An important geomorphologic and hydrologic question for flood risk analysis is whether the older surface has evolved out of the flood zone or whether it simply has not been flooded for 15,000 years because random channel migration across the fan took the locus of flooding and sedimentation far from the site for that length of time.

If, for example, the active sedimentation zone is now migrating into the older surface through lateral bank erosion, or lies at an elevation only a few meters below that surface (i.e., within the range of flood stage), or if upstream of the site there is an opportunity for avulsions that could lead channels or sheetfloods across the older surface, then the older surface lies within the zone subject to alluvial fan flooding.

If, by contrast, channels have become entrenched during the past 15,000 years for one of the reasons given above, the elevation difference between the recently active sedimentation zone and the older surface may be greater than any flood or debris flow stage conceivable in the current regime of climate, hydrology, or land use in the source area. In this case, "overbank" flooding is not possible. Maximum conceivable flood heights, albeit approximate because of the assumptions that must be made about potential changes of bed elevation, are predictable through methods described by Burkham (1988) for floods and Whipple (1992) for debris flows. If field inspection reveals that the margin between the older and the recently active surfaces is armored with bouldery or cohesive sediment, and if the ages of trees along this margin as well as the aerial photographic record (now approximately 50 years old) indicate little or no migration of that boundary, then one's confidence is increased that lateral bank erosion does not threaten the site.

Such evidence is found, for example, along many channelized alluvial fans that have been entrenched and rendered less active by changes of climate and hydrology. Former debris flow activity, or large meltwater or monsoonal Pleistocene floods, have often built fans of coarse sediment that were trenched by smaller floods soon after the climate change and now have channels that can contain all the stream discharges conceivable under current environmental conditions and are lined with sediment too resistant to be eroded rapidly. More equivocal conditions occur along channels that have undergone less extreme changes, but photographic and dendrochronological evidence indicates that channels on some, even fine-grained, fans do not aggressively undermine their root-reinforced banks, presumably because bank scour is not required by the combination of flood lows and sediment balance in the reach. Finally, even if the older surface is judged to be free from the risk of overbank inundation and undermining, the channel system upfan needs to be systematically surveyed for the potential of an avulsion that would lead flows across the older surface. This would require a combination of geomorphologic survey (for low banks, elevated channels, zones of sediment accumulation, bank erosion and channel shifting, and topographic lows on the fan surface) and calculations of the heights of conceivable floods and debris flows. On many alluvial fans, however, there is a relatively clear separation between older, higher, stable parts of fans, and channels or narrow inset floors trenched into the fan deposits.

The kinds of field evidence that can be used for making such determinations are reviewed in Chapter 3, and valuable sources of information are listed in Appendix B. It is unfortunate that the problem of interpreting the significance of different ages of fan surface for alluvial fan flooding risk is often not reduced by a simple indicator or numerical index. However, with a systematic approach such as the example given above (identify the risk of overbank inundation, lateral bank erosion, and inundation by avulsion from upstream), most field situations can be classified as a basis for both regulation and choice of a method for estimating flood risk, as required by FEMA. Complicated situations, however, will tax any approach.

REPORTS OF FLOODING ON ALLUVIAL FANS

Documented accounts of flooding confirm that active fans function primarily as a terminus for water and debris, while relict fans and streams function to convey water and sediment. Appendix A summarizes accounts of flooding at 29 individual or groups of alluvial fans to illustrate a wide variety of flood conditions on streamflow, debris flow, and composite fans. Most of the alluvial fans are in the arid southwestern United States, and a few are in humid areas throughout the world. The accounts are for observed floods and consist of direct and indirect measurements and observations of flood characteristics and geomorphologic interpretation of flood remnants. Only a few maps of the extent of flooding are available, and direct measurements of flow depths and velocities on fans are rare. Opportunities for systematic collection of flood data on alluvial fans are also rare.

These accounts demonstrate the complex nature of flooding on alluvial fans, ranging from flooding in stable channels to massive deposition of debris in urbanized areas. Accounts by different observers of a particular flood on a particular fan can depict different characteristics. This is because (1) observations from a single vantage point of the widespread and variable nature of fan flooding are limited and not desirable and (2) local conditions change during a flood because of the rearrangement of the channel system by remobilization of deposited debris, as described in the previous section. Also, the flood characteristics on adjacent or nearby fans resulting from the same storm commonly are different because of local differences in rainstorm intensity, sediment availability in the source areas, and the respective differences in recent flow path history.

Although flooding accounts depict varied properties in time and space, they show certain distinctive properties when considered collectively. High flood flow velocities were reported for about half of the sites listed in Appendix A. Supercritical flow velocities for water floods are also suggested at other sites by the steep fan slopes and low hydraulic roughness associated with fine-textured surface sediments and simple channel geometries. Flash floods (short-lived floods with a short time to peak) are reported at eight sites, including accounts of translatory waves at the Chicago Creek, British Columbia; Horseshoe Park, Colorado; and Montrose, California, alluvial fans. At the Cottonwood Canyon, Magnesia Spring Canyon, and Montrose alluvial fans in California the reported combination of high flow velocities, flashy flow, unstable channels, and movement of flow paths produced an especially serious hazard that resulted in loss of life at two of the fans.

Sheetflooding and distributary flow also are typical of alluvial fan flooding. Unstable channel boundaries are common, while stable channel beds or banks are less frequent, as suggested by accounts at only 5 of the sites studied. Flow path movement was reported for 8 sites, no movement for 13 sites, and either no mention of movement or absence of movement at the remaining 8 sites. A possible reason for the lack of flow path movement and disruption of the surface of some of the active fans is the short duration of the floods in concert with a limited supply of unconsolidated medium to coarse material from the drainage basin. Typical distributary channels appear to scour and fill; however, stable flow paths do occur during individual floods.

The Carefree and Wild Burro alluvial fans in Arizona are examples of less active or inactive fans with some channel incision and a limited supply of medium-to-coarse sediment from the drainage basin. Most of the sediment delivered to the fans is carried by streamflow and is coarse sand and fine gravel. The basins are not very steep. For example, the basin of the Carefree fan is a pediment. On these fans the channels are slightly trenched below the surrounding fan surface and are lined with desert trees and shrubs. During recent major floods, there was no movement of the many distributary channels on either of these fans.

Flood hazards on the least active fan surfaces, such as the Carefree (see Chapter 4) and Wild Burro alluvial fans, can be deceptively small because of the small size of distributary channels and the rather long distances from mountainous drainage basins. For example, shallow floodflow unexpectedly entered several small defined channels during a major flood on the Saddle Mountain, Arizona, alluvial fan (USACE, 1993). A subsequent field investigation revealed no new channel formation on the fan. Rather, small distributary channels that in places appeared to be simple topographic lows conveyed overflow from the larger distributary channels or carried runoff from high-intensity rainfall directly on the fan surface. Many less active alluvial fans in the arid southwestern United States have developed soils with small, slightly incised stable flow paths that convey shallow high-velocity floodflow during major floods.

Flood control works such as levees and debris dams at several of the sites were partially or totally ineffective during major floods. For example, at Day and Deer creeks (Figure 2-2), Henderson Canyon (see Chapter 4), and Magnesia Spring Canyon alluvial fans in California, the flood control structures were overwhelmed, and floodwater followed original flow paths and fan topography at and below sites of structural failure. This suggests that (1) major flood control works are necessary to mitigate flood hazards on active alluvial fans, (2) predevelopment fan topography influences the location of major flooding even after fans are urbanized and minor flood control structures are in place, and (3) flood control works must be designed to address specific types of hazards and special design consideration should be given for areas where water can still reach after flood control structures are installed.

The small Glendora alluvial fan is highly urbanized, and the paths of floodwater for a flood in January 1969 were significantly controlled by the network of streets that cross the fan parallel and at right angles to the general direction of fan slope (Figure 2-7). The alluvial fans of Cottonwood Creek and Stuart and Crane Gulches are also highly urbanized, and the paths of floodwater are significantly controlled by the conveyance capacity and location of many streets. A FIRM showing alluvial fan flooding hazards based on the assumption of uniform risk, thereby ignoring the channels created by the streets, would be inappropriate for these urban areas.

The Horseshoe Park and Wadi Mikeimin alluvial fans each were formed during a single flood. The Horseshoe Park alluvial fan was formed by a catastrophic flood from a dam failure. The fan formed at a topographic break at the mouth of the Roaring River. Wadi Mikeimin also formed at a break in floodflow confinement at the mouth of a river. The Roaring and Mikeimin River fans were significantly modified by erosion following the fan-building floods.

Abnormally large volumes of sediment were produced by storm runoff from recently burned basins of the Montrose, Cottonwood Creek, and Wasatch front alluvial fans. The close proximity of these fan basins to urbanizing areas may result in a greater incidence of wildfires with associated debris flows and alluviation on urbanized fan surfaces.

Reported flood characteristics of the sample of fans include high-velocity water flows, debris flows, translatory waves, sheetflood, distributary flow, unstable and stable channel boundaries, movement of flow paths, stable flow paths, and alluviation on the unchanneled fan surface. The composite of these accounts of flooding shows a wide variability of processes and flood hazard in time and space, which places a premium on field inspection and interpretation of concrete evidence from each alluvial fan before a determination is made of flood risk.

REFERENCES

Beaty, C. B. 1963. Origin of alluvial fans, White Mountains, California and Nevada. Annals of the Association of American Geographers 53:516-535.

Benda, L., and T. Dunne. 1987. Sediment Routing by Debris Flow. Publication. 165. Oxford, England: International Association of Hydrological Sciences.

Brunner, G. W. 1992. Numerical simulation of mudflows from hypothetical failures of the Castle Lake debris blockage near Mount St. Helens, Washington, Appendix A: Derivation of the equation to calculated the ultimate bulking factor. In Program for Steep Streams Workshop, Seattle, Washington, October 27-29, 1992. Portland, Ore.: U.S. Army Corps of Engineers.

Bull, W. B. 1977. The alluvial fan environment. Progress in Physical Geography 1(2):222-270.

Bull, W. B. 1991. Geomorphic Responses to Climate Change. New York: Oxford University Press.

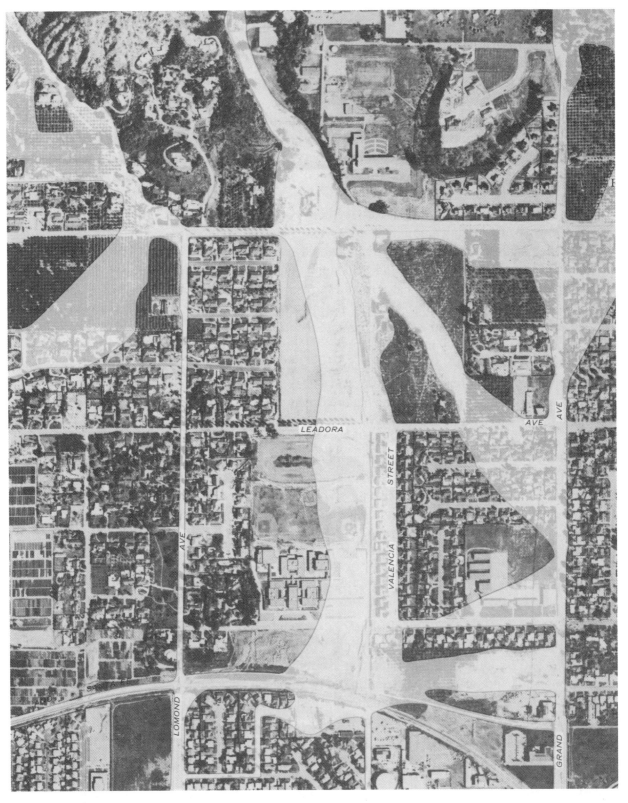

FIGURE 2-7 In urbanized settings, such as on the Glendora fan in California, streets can influence the path of floods. This area flooded in January 1969. SOURCE: Giessner and Price (1971).

Burkham, D. E. 1988. Methods for delineating flood-prone areas in the Great Basin of Nevada and adjacent states. U.S. Geological Survey Water-Supply Paper 2316. Reston, Va.: U.S. Geological Survey.

Dunne, T. 1991. Stochastic aspects of the relations between climate, hydrology, and landform evolution. Transactions, Japanese Geomorphological Union 12:1-24.

Fairchild, L. H. 1985. Lahars at Mount St. Helens. Ph.D. dissertation, University of Washington, Seattle.

French, R. H. 1995. Estimating the depth and length of sediment deposition at slope transitions on alluvial fans during flood events. Journal of Soil and Water Conservation 50(5):521-522.

Geissner, F. W., and M. Price. 1971. Flood of January 1969 Near Azusa and Glendora, California. U.S. Geological Survey Hydrologic Atlas HA 424. Reston, Va.: U.S. Geological Survey.

Hjalmarson, H. W. 1994. Potential flood hazards and hydraulic characteristics of distributary-flow areas in Maricopa County, Arizona. U.S. Geological Survey Water-Resources Investigations Report 93-4169. Reston, Va.: U.S. Geological Survey.

Hooke, R. L. 1967. Processes on arid-region alluvial fans. Journal of Geology 75:438-460.

Hooke, R. L., and R. I. Dorn. 1992. Segmentation of alluvial fans in Death Valley, California, new insights from surface exposure dating. Earth Surface Processes and Landforms 17:557-574

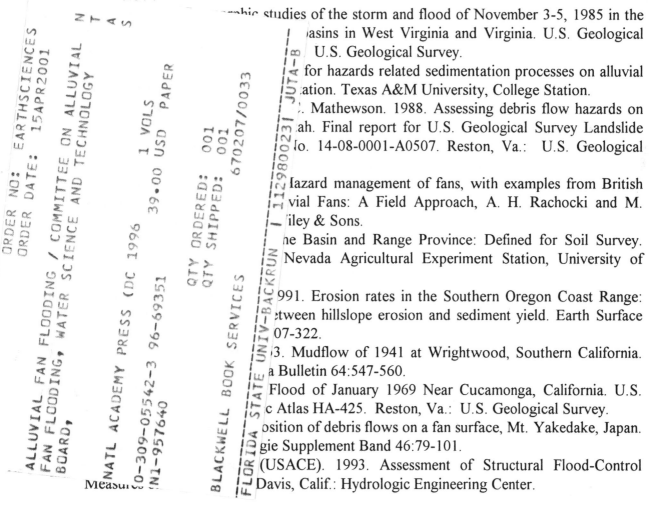

⸻hic studies of the storm and flood of November 3-5, 1985 in the

⸻ asins in West Virginia and Virginia. U.S. Geological

⸻ U.S. Geological Survey.

⸻ for hazards related sedimentation processes on alluvial

⸻ ation. Texas A&M University, College Station.

⸻. Mathewson. 1988. Assessing debris flow hazards on

⸻ ah. Final report for U.S. Geological Survey Landslide

⸻ lo. 14-08-0001-A0507. Reston, Va.: U.S. Geological

⸻ lazard management of fans, with examples from British

⸻ vial Fans: A Field Approach, A. H. Rachocki and M.

⸻ iley & Sons.

⸻ he Basin and Range Province: Defined for Soil Survey.

⸻ Nevada Agricultural Experiment Station, University of

⸻ 991. Erosion rates in the Southern Oregon Coast Range:

⸻ etween hillslope erosion and sediment yield. Earth Surface

⸻ 07-322.

⸻ 3. Mudflow of 1941 at Wrightwood, Southern California.

⸻ a Bulletin 64:547-560.

⸻ Flood of January 1969 Near Cucamonga, California. U.S.

⸻ c Atlas HA-425. Reston, Va.: U.S. Geological Survey.

⸻ osition of debris flows on a fan surface, Mt. Yakedake, Japan.

⸻ gie Supplement Band 46:79-101.

⸻ (USACE). 1993. Assessment of Structural Flood-Control

⸻ Davis, Calif.: Hydrologic Engineering Center.

Whipple, K. X. 1992. Predicting debris-flow runout and deposition on fans: the importance of the flow hydrograph, erosion, debris-flow and environment in mountain regions. Pp. 337-345 in Proceedings of the International Association of Hydrologic Sciences Symposium, D. Walling, T. H. Davies, and B. Hasholt, eds. IAHS Publication 209. Oxford, England: IAHS Press.

Whipple, K. X., and T. Dunne. 1992. The influence of debris-flow rheology on fan morphology, Owens Valley, California. Geological Society of America Bulletin 104:887-900.

3

Indicators for Characterizing Alluvial Fans
and Alluvial Fan Flooding

Alluvial fans and alluvial fan floods show great variability in climate, fan history, rates and styles of tectonism, source area lithology, vegetation, and land use. For this reason, it is essential that any investigation of alluvial fan flooding include careful examination of the specific fan for which information is needed. The committee recognizes that the extent of site-specific examination will be constrained by factors such as the amount of time and money allotted to the project, the tools available to the investigator, and the investigator's experience. As discussed in this chapter, however, much information can be gleaned from topographic and soil maps, as well as aerial photographs. Nevertheless, it is essential to do at least one field inspection of the fan that involves walking across its surfaces and along its channels. In general, the more fieldwork done, the better the understanding of the processes of flooding on the fan of interest.

According to the definition presented in Chapter 1, for regulatory purposes alluvial fan flooding is a flood hazard that on active parts of alluvial fans has a 1 percent chance of occurrence, and it is identified by flow path uncertainty and deposition and erosion below the hydrographic apex. The criteria used to assess whether an area is, or is not, subject to alluvial fan flooding must determine whether the flooding occurs on an alluvial fan and whether it is characterized by deposition, erosion, and flow path uncertainty below a hydrographic apex. For these reasons, the process of determining whether or not alluvial fan flooding can occur at a given location, and of defining the spatial extent of the 100-year flood, are divided into three stages:

1. Recognizing and characterizing alluvial fan landforms.
2. Defining the nature of the alluvial fan environment and identifying areas of active erosion, deposition, and flooding (as well as inactive areas).
3. Defining and characterizing areas on active parts of alluvial fans that are subject to a 1 percent chance of occurrence (the 100-year flood), the FEMA mandate.

Progression through each of these stages results in a procedure that narrows the problem to smaller and smaller areas of uncertainty (Figure 3-1). In Stage 1, the landform on which flooding occurs must be characterized. If the location of interest is an alluvial fan, then the user progresses to Stage 2, in which those parts of the alluvial fan that are active and inactive are identified. The term *active* means those locations where flooding, erosion, and deposition have

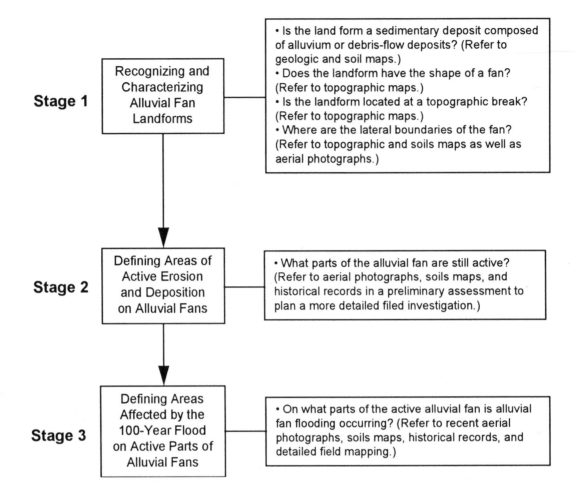

FIGURE 3-1 Three stages in the procedure to determine areas susceptible to alluvial fan flooding.

occurred on the fan in relatively recent time, and probably will continue to occur on that part of the alluvial fan. Those parts of the fan that have been active in relatively recent time can be identified depending on data availability for the site and money allotted to the project. (See Box 3-1.) Each active part of the alluvial fan also is characterized based on the dominant types of processes that result in flooding and sedimentation. Finally, in Stage 3, the evaluator determines where on those parts of the fan that are active the 100-year flood is possible. Progression through each of these stages will require the investigator to refer to a variety of maps, photographs, and other information sources (Table 3-1; also see Appendix B) and to do a significant amount of fieldwork to understand and characterize the alluvial fan flooding hazard.

BOX 3-1
"TIME" IN THE CONTEXT OF ALLUVIAL FAN FLOODING

It is not possible to give a precise definition to the phrase "relatively recent" as used in this report. Because of the variability among fans, the complexity of single fans, and the great range of ages present on fan surfaces, as well as tremendous diversity in the climate and geologic processes operating on fans, the committee is not able to specify a single time frame to use in defining whether or not a fan is active. Instead, such a judgment must be made on a site-specific basis.

At one extreme, the committee considers "recent" to be the past 10,000 years (the Holocene Epoch), which follows a particularly radical and widespread climate change at the end of the most recent Ice Age. At the other extreme, "recent" implies the time period over which there has been a relatively uniform range of environmental conditions that affect flood generation and channel behavior. Estimates of the probability of events occurring in the relevant, near-term future are based on the record of the "recent," homogeneous past. A problem exists, however, in that there often is no clear indication in most localities about how far back one should look in defining whether the record is relatively uniform, and this is in part why society is occasionally surprise by unforeseen flood hazards.

For purposes of the National Flood Insurance Program (NFIP), an arbitrary but reasonable decision was made to use, as a planning tool, the flood which has a 1 percent chance of occurring in any one year ("the 100-year flood"), which is usually a generally destructive flood in most areas of human settlement. Thus, the engineering perspective involves a timescale (a century) over which structures are typically designed to survive, while the geological perspective involves a longer time period with a greater range of geologic processes and environmental variability. The engineering perspective focuses on the regulatory requirements imposed by the NFIP; the geologic perspective focuses on geologic process. Both perspectives are important to understanding alluvial fan flooding. Examples of various attempts to determine ages of fans and fan components can be explored in depth in the following references:

Bull, W. B. 1964. Geomorphology of segmented alluvial fans in western Fresno County, California. U.S. Geological Survey Professional Paper 352-E. Reston, Va.: U.S. Geological Survey.

Bull, W. B. 1968. Alluvial fans. Journal of Geologic Education 17(3):101-106.

Bull, W. B. 1977. The alluvial fan environment. Progress in Physical Geography 1:222-270.

Kellerhals, R., and M. Church. 1990. Hazard management on fans, with examples from British Columbia. In Alluvial Fans: A Field Approach. New York: John Wiley & Sons.

Lecce, S. A. 1990. The alluvial fan problem. In Alluvial Fans: A Field Approach. New York: John Wiley & Sons.

Machette, M. N. 1985. Calcic soils of the southwestern United States. Geological Society of America Special Paper 203. Boulder, Colo.: The Geological Society of America.

Markewich, H. W., and S. C. Cooper. 1991. One perspective on spatial variability in geologic mapping. In Spatial Variabilities of Soils and Landforms, M. J. Mausbach and P. J. Wilding, eds. Soil Society of America Special Publication, no. 28.

Mausbach, M. J., and P. J. Wilding. 1991. Spatial Variability of Soils and Landforms. Soil Society of America Special Publication, no. 28.

Rhoads, B. L. 1986. Flood hazard assessment for land-use planning near desert mountains. Environmental Management 10(1):97-106.

Zarn, B., and R. H. Davies. 1994. The significance of processes on alluvial fans to hazard assessment. Z. Geomorph. N. F. 38:487-500.

STAGE 1: RECOGNIZING AND CHARACTERIZING ALLUVIAL FAN LANDFORMS

Determining Whether or Not a Landform Is an Alluvial Fan

The committee's definition of alluvial fan flooding specifically states that it occurs on alluvial fans. As a consequence, the first step in application of the definition is analysis of the area being considered for possible alluvial fan flooding. If this area does not meet the criteria for the definition of an alluvial fan, then it does not qualify for consideration of alluvial fan flooding. The committee defines an alluvial fan as *"a sedimentary deposit located at a topographic break such as the base of a mountain front, escarpment, or valley side, that is composed of streamflow and/or debris flow sediments and which has the shape of a fan, either fully or partially extended."* These characteristics can be categorized as composition, morphology, and location, as follows.

Composition

Alluvial fans are landforms constructed from deposits of alluvial sediments or debris flow materials.

To meet the criteria in the committee's definition of an alluvial fan, the landform of interest must be a sedimentary deposit, an accumulation of loose, unconsolidated to weakly consolidated sediments. In the following text, we use the term "alluvium" to refer to sediments transported by both streams and debris flows, but we emphasize that this is a grammatical convenience. On a particular fan, the distinction between these two forms of sediment transport is critical to a correct interpretation of the flood hazard.

Most sediments deposited during Quaternary time (2 million years ago to the present) still are loose and unconsolidated, as the processes of diagenesis that result in compaction, cementation, and lithification require millions of years to transform sediment to sedimentary rock. As a consequence, geologic maps commonly have a unit labeled "Qal" that conventionally is mapped in yellow and represents Quaternary alluvium. Determining whether or not a landform is an alluvial sedimentary deposit might be as simple as checking a published geologic map to see if the underlying material is mapped as alluvium. If a geologic map is not available, the user can check Natural Resources Conservation Service (NRCS) soil maps or drilling and logging records from water wells. If none of these sources is available, field reconnaissance can be done to determine whether or not the landform consists of alluvial sediments.

Morphology

Alluvial fans are landforms that have the shape of a fan, either partly or fully extended.

To meet the criteria in the committee's definition of an alluvial fan, the landform of interest must have the shape of a fan, either partly or fully extended. Flow paths radiate outward to the perimeter of the fan. This criterion can be assessed with topographic maps. For example, in Figure 3-2a the landform downstream from the Lawton Ranch, Montana, has the shape of a fan that is nearly fully extended. This landform is known as the Cedar Creek alluvial fan and is a classic example of a fan with nearly ideal morphology.

Location

Alluvial fan landforms are located at a topographic break.

To meet the criteria in the committee's definition of an alluvial fan, the landform of interest must be located at a topographic break where long-term channel migration and sediment accumulation become markedly less confined than upstream of the break. This locus of increased channel migration and sedimentation is referred to as the alluvial fan *topographic apex*. Figure 3-2 shows that the Cedar Creek alluvial fan begins at a topographic break, which in this case is a slightly embayed mountain front. As Cedar Creek exits its narrow bedrock canyon, it becomes less confined and is able to migrate more freely. Less confinement can lead to greater channel widths and smaller channel depths. As a result, the occurrence of deposition increases, and flow paths become more unstable.

Defining the Boundaries of an Alluvial Fan

Where are the toe and lateral boundaries of the alluvial fan?

Toe

The distal terminus, or toe, of an alluvial fan commonly is defined by

- a stream that intersects the fan and transports deposits away from the fan,
- a playa lake,
- an alluvial plain, or
- smoother, gentler slopes of the piedmont plain.

TABLE 3-1 Data Sources for Information on Alluvial Fans

Agency or Source	Source Number[a]	Topography	Surface Features	Land Cover	Land Use	Remotely Sensed	Aerial Photography
U.S. Department of the Interior							
Bureau of Land Management	1					☐	☐
National Park Service	2		☐		☐		
U.S. Geological Survey	3	☐	☐	☐	☐	☐	☐
U.S. Department of Agriculture							
Agricultural Stabilization and Conservation Service	4					☐	☐
Forest Service	5	☐					
Natural Resources Conservation Service	6		☐	☐	☐		
U.S. Department of Commerce National Ocean Service	7	☐					
U.S. Army Corps of Engineers	8						
Independent Federal Agencies							
Federal Emergency Management Agency	9						
Tennessee Valley Authority	10	☐	☐	☐	☐	☐	☐
National Archives and Records Administration	11					☐	☐
Library of Congress	12					☐	☐
Other agencies or sources:							
State geologists	13					☐	☐
State floodplain management agencies	14	☐				☐	☐
University libraries	15	☐	☐	☐	☐	☐	☐
County floodplain management agencies	16	☐				☐	☐
Long-time residents	17				☐		
Newspapers	18						
Technical journals	19						
University theses	20			☐	☐		

[a]See Appendix B.

Orthophoto-quads	Satellite	Hydrologic	Flood	Hydrography	Water data	Floodplain	Subsurface	Geology	Soils	Other
☐										
☐	☐	☐	☐	☐	☐	☐	☐	☐	☐	☐
		☐		☐						
		☐	☐	☐			☐		☐	
		☐	☐			☐				
		☐	☐			☐				
☐		☐		☐			☐	☐	☐	☐
							☐	☐		
		☐	☐				☐	☐		
		☐	☐		☐	☐				
			☐				☐	☐		☐
			☐		☐	☐				
			☐							
			☐							
			☐				☐	☐		☐
		☐	☐				☐	☐		☐

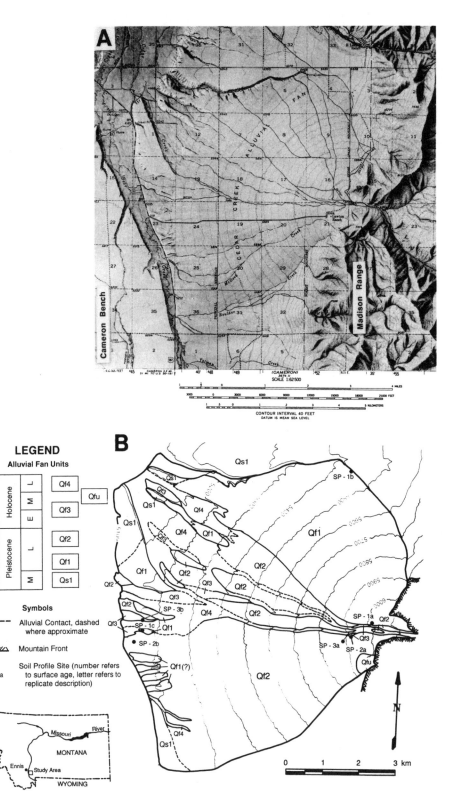

FIGURE 3-2 (a) Shaded relief map and (b) geologic map of Cedar Creek alluvial fan in Montana. SOURCE: Reprinted with permission from Ritter et al. (1993).

Such boundaries often can be identified on the basis of changes in the shapes of contour lines on topographic maps. For example, at the toe of a fan contour lines may become straighter or less concave when viewed downslope, although in the case of deeply dissected fans, contour lines may become more irregular and crenulated because of channel incision. The toe of the Cedar Creek[1] alluvial fan (Figure 3-2a) is defined by Bear Creek along the fan's western margin, and by the much larger valley floor of the Madison River into which Bear Creek flows along the fan's northwestern margin. Streams draining the northern part of the fan are more deeply incised because the Madison River valley floor forms a lower base level for erosion than its tributary valley floor along Bear Creek.

The toe of some alluvial fans in arid regions is indicated by a relative increase in the amount, size, and type of vegetation because ground water is closer to the surface there than on the upper parts of the fan. The toe of some alluvial fans in humid regions may be indicated by relatively less vegetation because the recent deposits are less fertile than older sediments. A general sense of vegetation types often is indicated on topographic maps.

Lateral Boundaries

Lateral boundaries of alluvial fans are the edges of deposited and reworked alluvial materials. The lateral boundary of a single alluvial fan typically is a trough, channel, or swale formed at the lateral limits of deposition. Crenulations in contour lines at fan boundary troughs can be observed along the margins of the Cedar Creek alluvial fan (Figure 3-2a).

Lateral boundaries of single alluvial fans commonly are distinct contacts between light-colored, freshly abraded, alluvial deposits and darker-colored, weathered deposits with well-developed soils on piedmont plains. Soils of active alluvial fans typically are less oxidized and lower in clay content than soils on older landforms. As a consequence, the younger soils generally are lighter colored and more friable. Color and texture changes often are pronounced on aerial photos or infrared remote sensing imagery. In areas with rock varnish formation,[2] the lighter surfaces of recent alluvial fan deposits in contact with undisturbed varnished surfaces of older deposits form a distinct boundary or contact that readily is distinguished by the relative darkness of the ground on aerial photographs and by on-the-ground inspection. Dark, undisturbed surfaces of rock varnish are found on old piedmont and valley deposits throughout the Basin and Range province of the western United States.

In the case of multiple fans that coalesce to form *bajadas*, where deposits and reworked material of adjacent alluvial fans merge, the boundaries between adjacent fans may be less distinct than those of individual fans adjacent to streams, rivers, or smooth piedmonts, but generally are

---

[1] The committee has not visited the Cedar Creek fan and inspected its surface and deposits. It is used as an example here because it has been studied intensively by prominent geomorphologists, and thus much information is available regarding it. In addition, it is a classic alluvial fan in shape and history.

[2] Rock varnish is a dark coating (from 2 to 500 microns thick) that forms on rocks at and near the Earth's surface as a result of mineral precipitation and eolian influx. The chemical composition of rock varnish typically is dominated by clay minerals and iron and/or manganese oxides and hydroxides, forming red and black varnishes, respectively. With time, the thickness of the coating increases if abrasion and burial of the rock surface do not occur. As a result, clastic sediments on alluvial fan surfaces that have been abandoned for long periods of time have much darker and thicker coatings of varnish than do younger deposits.

marked by a topographic trough or ridge. Although it is difficult to separate young deposits on one fan from similar age deposits on a coalescing fan, it sometimes is possible to distinguish them based on different source-basin rock types. For example, Bull (1963, 1964) defined fan boundaries in central California using contour maps, aerial photographs, and tests of the gypsum content of core hole samples. Bull found that the gypsum content of fan deposits derived from drainage basins underlain predominantly by clay-rich rocks was about five times that of fan deposits from drainage basins underlain predominantly by sandstone.

Boundaries of many alluvial fans are defined on 7.5-minute series orthophoto base maps by the NRCS. In the U.S. Southwest, typical NRCS soil series for alluvial fans include the Ramona, Soboba, Kinburn, and Anthony. Large areas where material from stream banks is freshly deposited and partly reworked during floods also are mapped, and smaller areas are identified as part of a particular series where the reworked material is located. For small alluvial fans less than about .8 km^2 (.3 mi^2), the detail of the mapped soil units on the 7.5-minute soil map series may not be sufficient to show many distributary channels and the fan boundaries. Soil maps used in conjunction with aerial photographs are an excellent means to define fan boundaries.

The nature and extent of alluvial fan flooding can be partially determined from published topographic, soils, and geologic maps and other sources of data. However, the committee emphasizes the importance of a field inspection by a qualified professional with experience and technical knowledge of geomorphology, slope stability, avalanche potential, flood hydraulics, flood hydrology, sedimentary facies, and alluvial fan processes. The general use of secondary information and the importance of field information is described in this chapter and in the examples described in Chapter 4.

STAGE 2: DEFINING THE NATURE OF THE ALLUVIAL FAN ENVIRONMENT AND IDENTIFYING THE LOCATION OF ACTIVE EROSION AND DEPOSITION

Most alluvial fans have parts that are active and parts that are inactive. Alluvial fan flooding occurs on active parts of alluvial fans.

In Stage 2, evidence is obtained that identifies areas of potential flooding. This step narrows the area of concern for Stage 3, which is the specification identification of the extent of the 100-year flood. Although alluvial fan flooding has occurred on all parts of an alluvial fan at some time in the geologic past in order to construct the landform itself, this does not mean that all parts are equally susceptible to alluvial fan flooding now. In fact, in most of the United States it is possible to identify parts of alluvial fans that were actively constructed during Pleistocene time (about 2 million to 10,000 years ago) and parts that have been active (i.e., flooded) in the Holocene (the past 10,000 years). The reason that this broad distinction generally is straightforward and simple in practice is that the two time periods were identified and defined on the basis of different climatic conditions. The Holocene epoch is a time of interglacial warm conditions, whereas the Pleistocene epoch was marked by repeated full glacial, cool conditions alternating with warm interglacials like that of the Holocene (Figure 3-3). During glacial times, ice masses expanded and advanced, evaporation was low, and in the dry western U.S. ground water tables and stream discharges were high relative to interglacial times. As a result of these climatic differences, flooding and sedimentation occurred at different rates and magnitudes during the

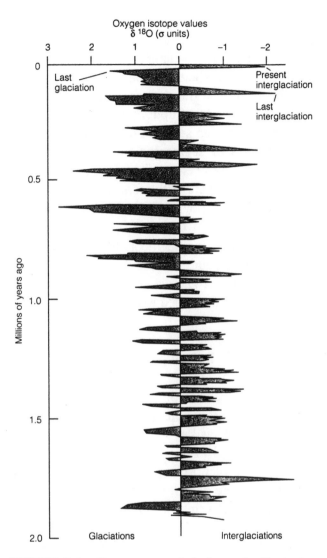

FIGURE 3-3 Quaternary period timescale illustrating oscillating climatic conditions from full glacial (cool) to interglacial (warm). SOURCE: Reprinted with permission from Skinner and Porter (1995).

Pleistocene and Holocene epochs. In many regions, the post-Pleistocene change of climate resulted in a reduction in the rate of sediment supply to fans, whether by streams or debris flows. As a result, the discharges presently available are able to move the sediment supplied on a lower slope than that formed during the Pleistocene, so fan-head incision is occurring on some fans.

One of the most common causes of the abandonment of large parts of an alluvial fan is a change in elevation of local streams. Elevation change can result from a change in climatic conditions or in rates of tectonism. Climatic change might result in a decrease in the size of large streams and/or lakes at the toe of the fan, as in the case of a change from braided, post-glacial meltwater streams to smaller, meandering streams incised into the braided gravel deposits.

A change in the rate of tectonic uplift along a mountain front also can result in abandonment of parts of alluvial fans. For example, a decrease in the rate of uplift at a mountain front relative to the fan could result in stream-channel downcutting at the mountain front/fan apex over a period of time. As a consequence, the upper part of the fan would become entrenched and the area of active alluvial fan deposition would shift downfan[3] (Figure 3-4). The opposite also can happen. In this case, an increase in the rate of uplift can result in rapid deposition at the fan head and development of a young untrenched fan segment overlying the older fan surfaces.

It is clear from the examples of segmented fans that only certain parts of the fan, or segments, are active at any one time. In the entrenched fan, the distal segment downstream of the hydrographic apex is typically the active part of the fan. In the untrenched fan, the segment of the fan proximal to the source area, at the topographic break, is typically the active part of the fan. These examples, however, are simplistic in that few fans have only one active segment that is clearly distinguished from an older, inactive segment. More typically, fans must be mapped to identify surfaces of different ages, from youngest to oldest.

To determine what parts of a fan are active and inactive, the investigator must examine the whole fan using indicators of activity as described here.

Defining Active

Because it is not possible to predict with zero uncertainty where the next flood will occur on an alluvial fan, we resort to the traditional method of the geologist and rely on the dictum "the past, as preserved in the geologic record, is key to understanding the present and to predicting the future." Using this reasoning, the geologist concludes that the area of deposition on an alluvial fan shifts with time, but the next episode of flooding is more likely to occur where the most recent deposits have been laid down than where deposits of greatest antiquity occur.

Once the planner has incorporated this basic philosophy into efforts to identify those parts of the fan that are active, the next step is to decide what time period will be used to define active. As a conservative standard, in some areas it is easy to separate the parts of a fan that formed more than 10,000 years ago from those parts that formed during the past 10,000 years, the Holocene epoch. In many places in the southwestern United States, it is less easy to see a clear Pleistocene/Holocene change, but one can subdivide Holocene deposits on a much finer timescale. In areas with long records of flooding and aerial photographs that date back to about 60 years, it is possible to identify areas of historic flooding within the past 100 years.

The term *active* refers to that portion of a fan where flooding, deposition, and erosion are possible. If flooding and deposition have occurred on a part of a fan in the past 100 years, clearly that part of the fan is active. If flooding and deposition have occurred in the past 1,000 years, that part of the fan can be considered to be active. However, it becomes more difficult to determine whether or not a part of the fan that has not experienced sedimentation for (say) more

---

[3] Note that the Cedar Creek fan is similar to the entrenched fan shown in Figure 3-4, except that its toe is bounded by a river that transports sediment away from the distal part of the fan. As a result, a young fan segment is not forming or preserved at the fan toe.

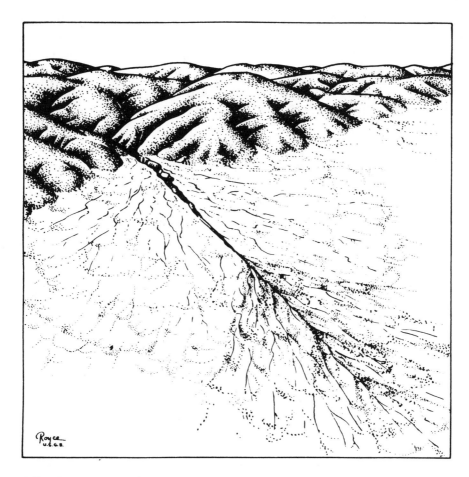

FIGURE 3-4 Entrenched alluvial fan with deposition occurring at the distal part of the fan. SOURCE: Reprinted with permission from Bull (1977).

than 1,000 years really is active, that is, that there is some likelihood of flooding and sedimentation under the present climate. A systematic approach to the problem of estimating whether a place may be inundated after a long period of inactivity is suggested in Chapter 2, the section "Change Over Time," which deals with processes by which flooding and deposition can migrate across an alluvial fan to invade places that have long been outside the zone of active deposition, even in the current climate. Since there is no clear analytical technique for making such projections, estimates of the probable spatial extent of inundation involve systematically applied judgment, and the combination of hydraulic computations and qualitative interpretations of geologic evidence concerning the recent history and probable future evolution of channel forms, as well as flooding and sedimentation processes. The problem becomes even more difficult when one considers the likelihood that the environmental conditions affecting the generation of floods or debris flows in the source area may not have remained constant in time.

Judging where flooding and deposition might occur is particularly important in cases where land use patterns are substantially altered by human activity (e.g., the Wasatch Range, Utah) or where recent decades have been marked by more intense storm patterns. As an example of the latter, storms in southern Arizona have changed in this century from moderate-sized summer events with sources in the south to much larger fall and winter events coming from the west-southwest (Pacific Ocean). Some of this change in storm conditions is attributed to the increasing frequency of El Niño climatic events over the past few decades. In such cases, it probably is prudent to define active as more than just those parts of the fan that have been the sites of flooding and deposition in the past 100 or 1,000 years.

Identifying Areas of Flooding and Deposition on the Active Part of an Alluvial Fan

It is important to identify both active and inactive parts of the fan, because this provides a map of where flooding can occur as well as where it probably will not occur.

Preparing a Geomorphic Map of Different Age Fan Surfaces

Once a time period is chosen to represent the active part of a fan for the purpose of flood hazard assessment, the flood evaluator must determine which deposits are less than the chosen age. A simple place to start is to examine the historical record of flooding and sedimentation. Aerial photographs from different years can be compared to identify sites of deposition that are less than about 60 years in age. If humans have lived in the area, historical deposits often contain relicts of human activity, such as pieces of machinery, bottle caps, lumber, and scraps of metal.

These deposits can be examined and described to gain a good understanding of the nature of fresh deposits for that alluvial fan. The flood evaluator can map different deposits, placing them in relative chronological order from youngest to oldest, as mapping of the entire fan progresses. On a surface of essentially continuous deposition, gradational relations are the rule.

The product of this part of the investigation should be a basic geomorphic map of the entire fan, with particular emphasis on the active parts of the fan. An example of such a geomorphic map is shown for the Cedar Creek fan (Figure 3-2b). This map divides alluvial fan surfaces into different age categories, from as old as middle Pleistocene to as young as late Holocene. Soil profiles were described at different sites, and weathering characteristics such as those described below were used to assess the relative age of each surface. In this example, only a small part of the total fan can be considered to be active.

Morphologic and Weathering Criteria Used to Prepare a Geomorphic Map

A variety of properties can be used to separate deposits of different ages. These include features such as fan surface morphology and sediment weathering characteristics.

The surface of a recent deposit typically is irregular, whereas older deposits generally are smoother. Fresh stream-laid deposits commonly have bar and swale topography, whereas fresh debris flow deposits might have sharply defined levees along lateral margins. With time, loose, unconsolidated flood deposits weather, and some fine-grained material is added to the deposit from eolian influx. As deposits weather, clast size becomes smaller, and edges of deposits from individual flood episodes become more subdued. The net result is that the morphology of older surfaces becomes increasingly more subtle as micro- and macro-relief features are worn down. An exception to this general rule occurs, however, as alluvial surfaces become so aged and smooth that runoff can collect and begin to incise the surface. In such cases, older alluvial fan surfaces are characterized by very smooth, dark surfaces dissected by narrow channels.

Weathering characteristics that have been used by many workers to determine relative ages of alluvial fan deposits include desert pavement,[4] rock varnish, B-horizon development in the soil profile, calcic-horizon development in the soil profile, and pitting and rilling of clasts (Bull, 1991; Cooke et al., 1993; Dorn, 1994). In general, each of these characteristics becomes more pronounced and better developed with time, although the rate of development is site-specific owing to the influence of factors such as climate, eolian influx, and parent rock type.

Weathering parameters such as desert pavement, rock varnish, calcic-horizon development, and pitting and rilling are more useful in arid and semiarid regions than in humid regions. The use of these parameters has been common in dry climates, however, because of the scarcity of datable organic material in alluvial deposits. In humid regions, such distinctions are often less obvious, but workers often have the additional benefit of organic material that can be dated with radiometric carbon methods for time periods in the range of interest. If organic matter can be obtained from shallow trenches in deposits on alluvial fans, the investigator need not rely on other more relative weathering parameters, although these can be used to supplement the radiometric age estimates.

In the Basin and Range province of Arizona, California, Nevada, and Utah, alluvial fans are common, and their deposits can be correlated from one fan to another on the basis of relative-age criteria associated with morphologic and weathering characteristics. Christenson and Purcell (1985) identified eight characteristics that are useful in separating alluvial fan deposits into three general age categories throughout the region (Figure 3-5): young (less than 10,000 to 15,000 years), intermediate (10,000 to 700,000 years), and old (greater than 500,000 years). Their young category can be considered to be the active surface if the fan is not incised. The broad regional similarities in these deposits appear to be the result of Quaternary climatic changes (Christenson and Purcell, 1985). The eight characteristics identified and their nature for each of the three age groups are given in Table 3-2. Bull (1991) noted similar regional correlations over an even broader area that includes parts of northern Mexico and New Mexico.

In the eastern Grand Canyon, Hereford et al. (1995) were able to map three alluvial fan surfaces that range in depositional age from about 1000 B.C. (2950 years B.P.) to the present. Like Christenson and Purcell (1985), they referred to these deposits as older, intermediate, and younger (Figure 3-6), but their ages are 1 to 2 orders of magnitude younger. Although these surfaces are much more finely discriminated than those described by Christenson and Purcell (1983), the authors still were able to identify significant differences in weathering characteristics of each age surface (Table 3-3). Also, because of the young ages of the deposits, the authors were able to use archeological features and radiocarbon dating to supplement the weathering-related age criteria (Figure 3-7). In this example, only areas of channelized debris flows and younger debris flows could be considered to be active.

---

[4] Desert pavements are surfaces of tightly packed gravel that armor, as well as rest on, a thin layer of silt, presumably formed by weathering of the gravel. They have not experienced fluvial sedimentation for a long time, as shown by the thick varnish coating the pebbles, the pronounced weathering beneath the silt layer, and the striking smoothness of the surface, caused by obliteration of the original relief by downwasting into depressions (Ritter et al. 1995, pp. 252-253).

FIGURE 3-5 Young, intermediate, and old alluvial fan deposits, Gila Mountains, southern Arizona. SOURCE: Reprinted with permission from Christenson and Purcell (1985).

Vegetation Criteria Used to Prepare a Geomorphic Map

Vegetation types often differ from an alluvial surface of one age to that of another. The reasons seem to be related to the texture and composition of the sediment, as well as to the abundance and availability of moisture in the sediments. For example, on a fresh alluvial deposit, incipient soils contain little organic carbon or clay. As a result, the soils are low in nutrients and have little water-holding capacity. Older deposits are more enriched in carbon and clay content and have higher water-holding capacities.

Use and interpretation of diagnostic vegetation, just like the use and interpretation of desert pavement, varnish, or soil properties (e.g., clay or carbonate content) must be specific to the individual fan in question. For example, some mesquite species are riparian, but others can live anywhere in diminutive form. Palo Verde are more lush along waterways, but also can live well away from streams.

TABLE 3-2 General Characteristics of Young (<10,000 to 15,000 years), Intermediate (10,000 to 700,000 years), and Old (>500,000 years) Alluvial Fan Deposits, Basin and Range Province, United States

Characteristic	Young	Intermediate	Old
Drainage pattern	Distributary; anastomosing or braided	Tributary; dendritic	Tributary; dendritic or parallel
Depth of incision	Less than 1 m	Variable (1 to 10 m)	Greater than 10 m
Fan surface morphology	Bar and channel	Variable, generally smooth and flat	Ridge and valley, most of surface slopes
Preservation of fan surface	Currently active	Incised, but well-preserved wide, flat divides	Basically destroyed, locally preserved on narrow divides
Desert pavement	None to weakly developed	None to strongly developed	None (surface destroyed) to strongly developed (surface preserved)
Desert varnish	None to weakly developed (most varnished clasts reworked from older surfaces or bedrock)	None to strongly developed	None (surface destroyed) to strongly developed (surface preserved)
B horizon	None to weakly developed	Weakly to strongly developed	None (surface destroyed) to strongly developed (surface preserved)
Calcic horizon	None to weakly developed, $CaCO_3$ disseminated throughout	Weakly to strongly developed	None, carbonate rubble on surface (surface destroyed) to strongly developed petrocalcic horizon (surface preserved)

SOURCE: Reprinted with permission from Christenson and Purcell (1985).

Plant type, as well as vegetation density and diversity, is associated with surface age. Some plant species are riparian (ironwood), others are xerophitic (cacti), and others are completely intolerant of moist soil (e.g., saguaro). Vegetation density and diversity are low (but not nonexistent) in streambeds, become most dense and diverse for intermediate-age surfaces (middle to late Holocene), and become less dense and less diverse for old ridge-and-ravine surfaces. Streamflow limits vegetation in the channels by scour and removal of plants and their root support systems, but promotes vegetation on low terraces by watering them with overbank flow and water infiltrated into the bed. Vegetation is limited on old surfaces because they receive only direct rain, are often erosional, and can be less fertile (carbonate soil cropping out at the surface, for example).

On the Tortolita piedmont, in Arizona, surface age and vegetation are related in the following manner, with the most dominant plant listed first (Pearthree et al., 1992):

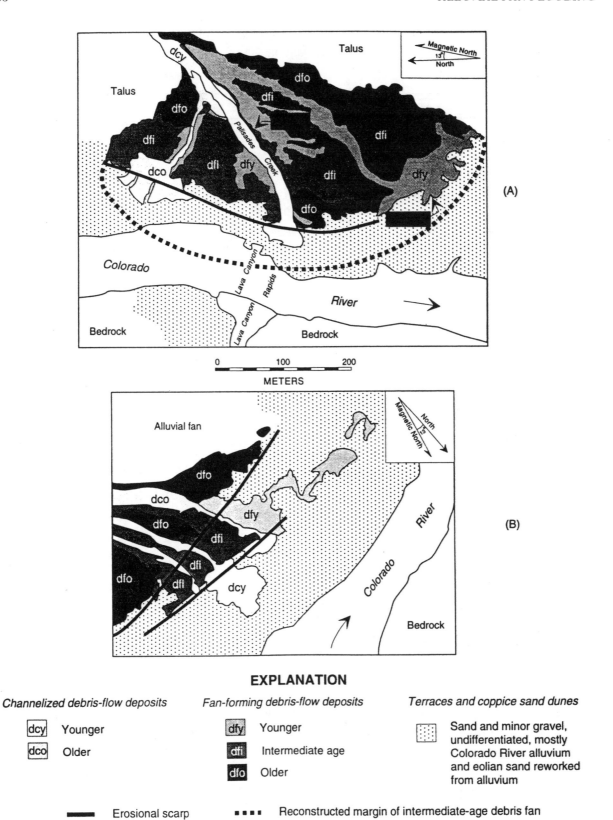

FIGURE 3-6 Geologic maps of debris flow fans in the eastern Grand Canyon. SOURCE: Hereford et al. (1995).

TABLE 3-3 Surface Weathering Characteristics of Clasts on Fan-Forming Debris Flows, Eastern Grand Canyon

Characteristic	Debris Flow Age Category		
	Younger	Intermediate	Older
Carbonate coatings, underside of clasts	None	Stage I, discontinuous, thin, <0.1 mm	Stage I, discontinuous to continuous, thin, <0.5 mm
Splitting, spalling, and granular disintegration of sandstone clasts	None	Slight	Common
Tafoni	None	Present	Present and well developed
Limestone-clast pitting[a]	None to incipient, <1 mm	Present, 1.47 to 4.04 mm	Present, 4.74 to 9.97 mm
Rilling of limestone clasts	None	None	Present on 5 percent of clasts
Rock varnish, sandstone clasts	Absent to incipient on 50 percent of clasts	Present on all 50 to 100 percent of clasts, brown to dark brown	Well developed on all clasts, dark brown to black

[a]Average depth of solution pits measured with a depth micrometer; number of individual measurements = 1,156 and 1,301 of intermediate age and older, respectively. Measurements made on the surfaces of 149 and 96 intermediate-age and older clasts. SOURCE: Hereford et al. (1995).

Late Holocene	Ironwood, grasses (after rainy season), Palo Verde, mesquite, bushes
Early Holocene	Palo Verde, ironwood, cholla, bushes
Latest Pleistocene	Bursage bushes, Palo Verde, cholla
Late Pleistocene (100 ka)	Saguaro, cholla, Palo Verde, bursage

Because recent deposits are likely to be within a zone of frequent flooding, it is unlikely that mature vegetation will occur on historical deposits. It sometimes is possible to find evidence of flood damage on vegetation, thus providing a clear means of identifying parts of fans that recently have been active.

Types of Alluvial Fan Flooding

Alluvial fan flooding, as described in the committee's definition, is characterized by flow path uncertainty below the hydrographic apex and caused by abrupt deposition of sediment, proximity to a sediment or a debris flow source area, sufficient energy to carry coarse sediment at shallow flow depths, and the absence of topographic confinement which may allow higher flows to initiate a new, distinct flow path of uncertain direction. Although such flooding occurs on the active part of an alluvial fan, the fact that an area is defined as active does not mandate that it also

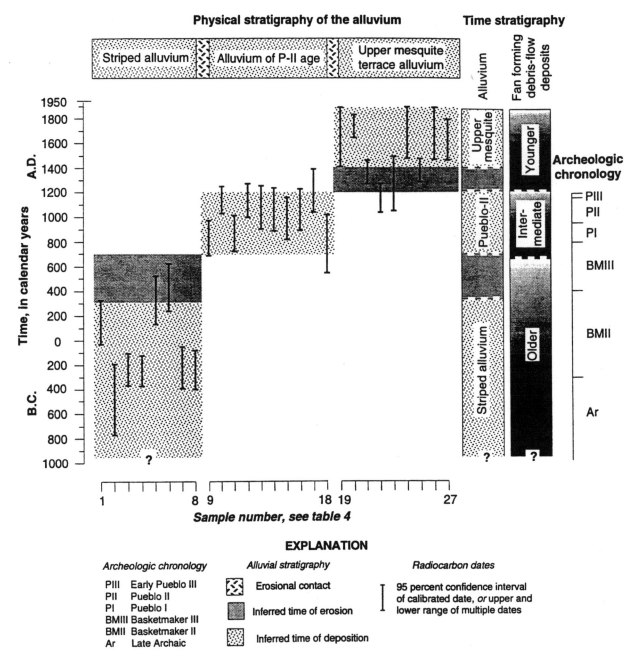

FIGURE 3-7 Time stratigraphy, physical stratigraphy, and archeological chronology, eastern Grand Canyon. SOURCE: Hereford et al. (1995).

is subject to 100-year alluvial fan flooding. Riverine flooding also occurs along the channels of many alluvial fans, especially those that are deeply incised. Even though flood hazards happen to be on alluvial fan landforms, they should be dealt with by FEMA under the guidelines established for river floodplains.

Identifying those parts of the active part of alluvial fans that are susceptible to alluvial fan flooding also requires examination of the types of flooding as recorded by flood deposits. The final result of such investigation should be a map of areas of different types of flood hazards. This map is not the same as a Flood Insurance Rate Map, which is a map of different flood hazard zones. The map recommended here is one that delineates the boundaries of areas of all types of flood hazards that occur on the active parts of the alluvial fan. Such a map requires the identification of locations where flow paths are uncertain, where erosion and deposition are likely to occur, where channels with confined flow exist, where channel avulsions have occurred or might occur, where sheet flow occurs, where debris flows occur, and where channelized stream flow with overbank flooding occurs. Mapping similar to that recommended here has been done, for example, for the Alberta Creek fan in Canada (Kellerhals and Church, 1990). Such maps can be included in flood insurance studies.

Defining Flooding Along Stable Channels

It is not uncommon for active parts of fans to contain stable channels that will not be susceptible to alluvial fan flooding. These channels might become unstable in the downstream direction, as in the case of entrenched alluvial fans. On the other hand, unstable channels can become stable in the downstream direction, as in the case of the dissected toe of the Cedar Creek alluvial fan shown earlier (Figure 3-2a). FEMA maps of alluvial fans should strive to indicate those channels susceptible to riverine flooding as well as those areas prone to alluvial fan flooding.

Identifying Areas Where Sheetflow Deposition Occurs

Some parts of alluvial fans are characterized by sheetflow, which is the flow of water as broad sheets that are completely unconfined by any channel boundaries. Sheetflow might occur where flow departs from a confined channel and no new channel is formed. It might also occur where several shallow, distributary channels join together near the toe of a fan and the gradient of the fan is so low that the flows merge into a broad sheet. Because such sheetflows can carry high concentrations of sediment in shallow water and follow unpredictable flow paths, they can be classified as alluvial fan flooding processes if they occur on alluvial fans. Sheetflow generally occurs on downslope parts of fans, where channel depths are low and the boundaries of channels become indiscernible. They are also more common at distal locations because of the likelihood of fine-grained sediments and shallow ground water; during prolonged rainfall, the ground can become saturated, resulting in extensive sheetflooding as runoff arrives from upslope. Fine-grained sediments can aggravate the likelihood of sheetflow because some clay minerals swell when wet, forming an impermeable surface at the beginning of a rainstorm.

Identifying Areas Where Debris Flow Deposition Occurs

Some parts of alluvial fans are characterized by debris flows. Debris flows pose hazards that are very different from those of sheetflows or water flows in channels (see Chapter 2). Identifying those parts of alluvial fans where debris flow deposition might occur requires the examination of deposits from past flows. Debris flow deposits can be distinguished from fluvial deposits by differences in morphology, depositional relief, stratigraphy, and clast fabric (Figure 3-8; Table 3-4). Exposures in channel banks can be examined and can be supplemented with shallow trenches in different deposits. In an example of a channel bank exposure described by Hereford et al. (1995) in the eastern Grand Canyon, debris flow deposits are interbedded with streamflow gravels, but can be distinguished by the differences in stratigraphy and clast fabric (Figure 3-9).

STAGE 3: DEFINING AND CHARACTERIZING AREAS OF 100-YEAR ALLUVIAL FAN FLOODING

For FEMA to carry out the mandates of the National Flood Insurance Program (NFIP), areas that are subject to flooding during a 100-year flood—that is, areas subject to a 1 percent chance of flooding in any year—must be identified. The two previous sections described methods of identifying landforms subject to alluvial fan flooding and the active portions of the fan that are subject to flooding. But identification of possible hazard areas is only the first step. The third step, one that is critical for floodplain managers and regulators, is to determine the severity and to delineate the extent of the 100-year flood, that is, the area exposed to a 1 percent risk of flooding in any given year. Although field work and study of aerial photographs and topographic maps are essential for carrying out the three stages necessary to identify alluvial fans and stable and unstable components of fans, the three-stage analysis can be quantified by the use of hydrologic methods. Although it is beyond this committee's scope and resources to explore in detail the numerous methods that have been developed to evaluate flood hazards, it is appropriate to give a general overview of the methods available to delineate the actual flood hazards on a fan. Thus this section briefly addresses the techniques, the types of analysis, and the appropriate perspectives that may be of assistance in the delineation of the hazards on alluvial fans and explores their potential for assisting FEMA in its mapping responsibilities. This discussion is not intended to be a complete exploration of all the methodologies that have been developed for hydraulic analyses, but rather it is a general introduction to several methods currently in use. In the future FEMA might consider conducting a detailed review of these methods and how they are applied

The mapping of flood risks for purposes of the NFIP is based on the flooding that is likely to result from an event that has the probability of occurrence of 1 percent in any given year, an event more commonly known as the 100-year flood. Within relatively stable river systems, it has been a standard practice to delineate the 100-year floodplain using a modeling technique based on the assumption that the flow is clear water and the hydraulic conditions are such that flow is gradually varied. In many instances, this technique also is used to model more dynamic systems with some acknowledgment of its limitations, because the areas of hazard within a river valley are usually apparent and confined to a geologic floodplain.

Areas subject to alluvial fan flooding often are not as readily apparent as those subject to riverine-type flooding. The physical characteristics of the fan shape also make the use of simplifying assumptions seem less logical and therefore less acceptable. Active alluvial fans are changeable, and erosion and deposition occur to some degree with most events. Inactive fans may also have flow paths that are unconfined and subject to uncertainty largely because of the numerous channel forks and joins.

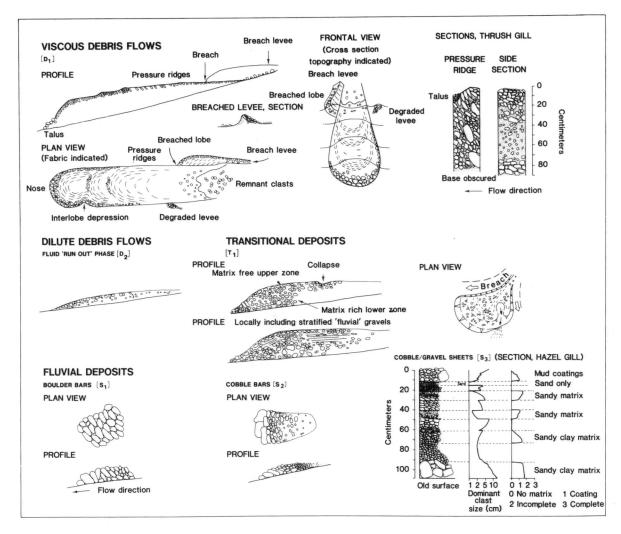

FIGURE 3-8 Morphologic and stratigraphic characteristics of different flow types developed from an example fan in England. SOURCE: Reprinted with permission from Wells and Harvey (1987).

When floodwater contains a significant amount of sediment or the flooded area is subject to scour and deposition, the flow behavior becomes less predictable. High concentrations of sediment and debris in flowing water can cause it to behave differently than clear water flows. Some of these differences, such as the unit weight, are quantitative in nature. Other differences, such as the vertical velocity distribution for a debris flow, display qualitative differences when compared to clear water.

TABLE 3-4 Summary of Morphologic and Sedimentologic Field Criteria for Distinguishing Facies Types

Facies Type	Morphology	Depositional Relief (m)	Texture and Stratification	Characteristics of Clast Fabric
Debris flow (D1)	Lobate to digitate; narrow; steep front and flanks; flat tops with low relief pressure ridges	High (0.8-1.5)	Matrix-rich (muddy); matrix-supported clasts; poorly sorted; $b$max range 80-210 mm; stratification absent	Elongate clasts oriented parallel to flow boundary forming a push fabric
Dilute debris flow (D2)	Thin lobate; broad, flat top; gentle lobe fronts and flanks	Moderate (0.3-0.5)	Matrix-rich; matrix supported clasts; poorly sorted; $b$max range 60-230 mm; stratification absent	None observed
Transitional flow deposits (T1)	Stacked lobes; broad small superimposed mounds; small collapse depressions	High (0.5-1.5)	Clast support with no matrix in upper few centimeters; matrix (sandy) increases with depth $b$max typically <180 m; moderately sorted; stratification present	Collapse packing
Fluvial boulder bar and lobes (S1)	Linear bars to transverse lobes	Moderate to high (0.5-0.8)	No matrix; clast support; front-to-tail sorting; $b$max typically >200 mm	Imbrication
Fluvial longitudinal bar (S2)	Linear bars	Moderate (0.2-0.5)	Clast support; matrix (sandy) increases with depth; market front-to-tail sorting; more poorly sorted than type S1; $b$max typically <120 mm	Strong imbrication
Fluvial sheet deposits (S3)	Broad and flat; some fan-shaped; subdued bar and swale forms	Low (-0.1)	Clast support; little matrix (sandy); well stratified; normal grading in some strata; moderate sorting in each stratum; $b$max typically <100 mm	Weak imbrication

SOURCE: Reprinted with permission from Wells and Harvey (1987).

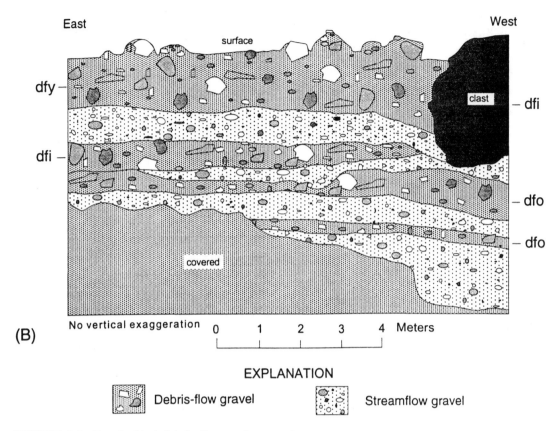

FIGURE 3-9 Interbedded debris flow and streamflow gravel, eastern Grand Canyon. SOURCE: Hereford et al. (1995).

AVAILABLE METHODS OF ANALYSIS

To investigate flood hazards, there are three general categories of interest: clear water flows that can be analyzed with traditional hydraulic methods, hyperconcentrated sediment flows that can be analyzed to a great extent by sediment transport theory, and debris flows that can be assessed by various empirical methods such as the bulking factor, the Bingham model, and other methods.

Appendix 5 of FEMA 37, *Guidelines and Specifications for Study Contractors* (1995), describes a method for delineating the boundaries of flood hazards on a fan shaped surface. This method, however, is the cause of some confusion. The method considers the conditional probability of the occurrence of a flood with a given magnitude, taking a certain path through the spatial domain, and inundating a point of interest. The equation that allows one to apply this method is called the total probability equation. Its purpose is to compute, for example, when the combined probability of two events is equal to 0.01. The events can be the occurrence of a flood, the failure of a levee, the coincidence with a different flood, the chance that floodwaters take a certain flow path, and so on. The purpose of using this method is to account for uncertainty when it cannot be easily set aside.

The use of the total probability equation is not limited to alluvial fans, and it is used by other federal agencies in addition to FEMA (NRC, 1995). The method of solving the total probability equation proposed by Dawdy (1979) has been used in the preparation of several FIRMs in the western United States. This method assumes that all areas of the fan are subject to flooding and that there is a fixed relationship between flooding depth and discharge. These assumptions apply when there is absolute uncertainty regarding how floods will occur. The advantages of these assumptions are that they are reproducible, they lend themselves to uncomplicated regulatory implementation, and, in certain situations, they are the easiest assumptions to defend. FEMA has developed a computer program called FAN (FEMA, 1990) that incorporates these assumptions, and it provides this program to contractors charged to delineate alluvial fan flooding.

When it comes to mitigation and the implementation of floodplain management regulations, however, it may be appropriate to review the assumption of complete uncertainty. There may be historical flow paths that are preferred during small floods. From a mitigation perspective, it would make sense to reinforce these paths rather than ignore them. The current Flood Insurance Rate Maps (FIRMs) that have been prepared using the procedure recommended by FEMA are a statement of complete uncertainty. These FIRMs then are not necessarily useful to floodplain mangers and regulators (who are often unaware of the procedures followed to identify the hazard) to assist them in determining the hazards on a particular fan area.

Since the decision on how or whether to solve the total probability equation is usually made by the flood insurance study contractor, the safe, default assumption of complete uncertainty is typically embraced to save, among other things, time. Most of the alluvial fan areas examined by the committee, however, show obvious, preferred flow directions. Alternative solutions to the total probability equation can be applied to these areas, but the guidelines in FEMA 37 (1995) are not clear about this and suggest only that the default assumption should be a starting point. Furthermore, permission to deviate from this assumption must be obtained in writing from FEMA.

All flooding sources have uncertainty. There is an apparent contradiction between the existing definition of alluvial fan flooding, which is very inclusive, and the actual method being used to delineate the hazard, which is limited to fan-shaped landforms. Flood behavior is predictable within the expected range of uncertainty. When the uncertainty can no longer be set aside but must be dealt with directly to achieve a reasonable result, then the total probability equation becomes a useful method for delineating flood hazards. The applicability of the method, however, does not mean that an area is subject to alluvial fan flooding. It is merely a way of expressing uncertainty.

FEMA has not developed guidelines on the general solution of the total probability equation. The committee recommends consideration of the use of *Guidelines for Risk and Uncertainty Analysis in Water Resources Planning* (USACE, 1992) for specific guidelines on how to apply the method.

The principles of risk-based analysis (USACE, 1992) provide a framework for a more general and realistic way to identify areas subject to flooding with an annual probability of 1 percent. The degree of uncertainty associated with a prediction of a given flood scenario is assessed by bringing to bear evidence derived from geomorphologic and other studies (for example, an alluvial fan with a series of branching channels). Figure 3-10 shows a flow diagram for conducting an analysis of diverging channels by considering various scenarios. Figure 3-11

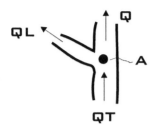

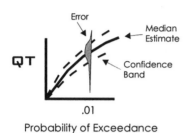

Flow path uncertainty exists at Point A. The total flow **QT** may split into components **QL** and **Q**.

1. Generate a likely annual maximum **QT**
2. Randomly sample the error and add to **QT**

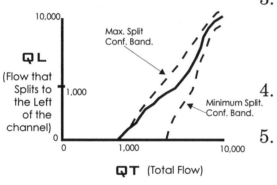

3. Where does the flow go? Develop a rating function describing flow path uncertainty.
4. Based on **QT** from step 2 determine flow split **QL**.
5. Randomly sample the error and add to **QL**. The remaining flow **Q = QT - QL**.

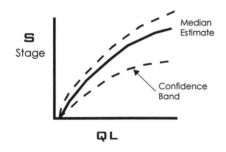

6. Route **Q** and **QL** down stream.
7. Based on a stage discharge relationship, estimate stage **S** along each flow path.
8. Randomly sample the error and add to **S**.
9. Repeat steps 3 through 9 for each flow split scenario

FIGURE 3-10 Analysis of flow path uncertainty considering possible scenarios.

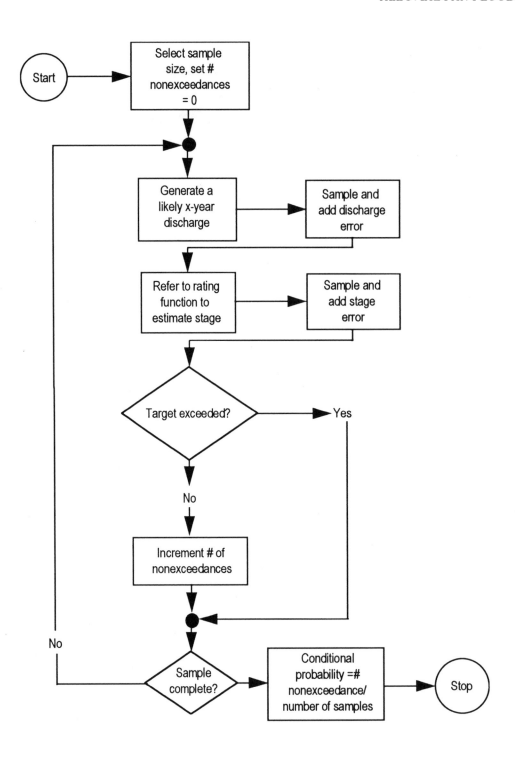

FIGURE 3-11 Conditional nonexceedance probability estimation with event sampling.

shows an example of estimating conditional nonexceedance probability using event sampling.

The broad spectrum of types of flooding that can occur and that have been observed on alluvial fans illustrates the futility of developing a "cookbook" method to apply to all fans in all geographic areas. Reviews of current research and discussions with local officials charged with regulating development in alluvial fan flooding areas indicate a prevailing preference for analysis of the flood hazards based on site-specific evaluations. The types of information that should be gathered include both geomorphic and process considerations. For example, some flow associated with channels meets the criteria for alluvial fan flooding, even though it occurs in channels in much the same way that riverine flow occurs in channels. The difference between the two is that in alluvial fan flooding the channels are likely to shift position with time and flows often abandon one channel to form another, resulting in much unpredictability regarding the locations of future flow paths. The following questions can provide guidelines to identify areas where flow paths are uncertain and flow is likely to leave confined channels to move in unpredictable directions:

- Where is there evidence of recent channel shifting?
- Where is there evidence of recent channel avulsion?
- Where is there evidence of the potential for channel avulsion?
- Where has channel geometry changed markedly in recent time?

Because alluvial fan flooding is associated with high rates of erosion, sediment transport, and deposition, it is common for such flows to shift position as sediment is dropped and forms obstructions to the flow. In some events, previous channels are completely blocked by deposits, and a new channel is formed. This process is known as avulsion and can be identified from aerial photos or field mapping by the presence of topographic lows (abandoned channels), the upstream parts of which filled with sediment. Areas of potential channel avulsion sometimes can be identified from construction of longitudinal and cross-fan profiles, because avulsion is likely to occur in places where sedimentation has raised the channel floor surface to a level that is nearly as high as the surrounding surface of the alluvial fan.

In addition, human modification of alluvial fan surfaces and urban development on alluvial fans have resulted in cases where human-made obstructions themselves have been the cause of alluvial fan flooding. For example, construction of culverts to divert water from one part of a fan to another sometimes results in rapid sedimentation downstream from the mouth of the culvert. The result can be that alluvial fan flooding then occurs in an area that might not have been mapped as susceptible to this type of flooding before human alteration of the landscape. Special attention is needed to identify areas where engineered works might aggravate or cause alluvial fan flooding during the time period designated as active by the investigator.

Specific steps that should be followed before undertaking any final delineation of alluvial fan flooding hazards include detailed office and field reviews of historical information and the evaluation of the present landform. Initial office procedures include the review of topographic maps and aerial photographs to determine the location and the morphology of the landform to determine whether it is a true alluvial fan. Other data that should be gathered early include historical maps and old photographs to document channel changes, changes in channel morphology, and the areas of the fan that may be classified as either active or inactive. Soil and geologic mappings should be examined to confirm the relative geologic age of fan deposits. Climatologic data and appropriate hydrologic analyses will be needed to determine the magnitude and frequency of flooding to be addressed. Aerial photographs and geologic information of the catchment area will provide indications of the amount of sediment and debris that can be delivered to the fan.

Field investigations by a trained observer should include gathering information on elevation differences across the fan and in a transverse direction if detailed topographic maps are not available. Vegetation types, soil characteristics, and the presence of desert varnish should be added to the office maps to confirm the active or inactive portions of the fan. Observations and measurements of channel conditions should be made to determine areas of possible avulsion. Detailed inspection of diffluences or abandoned channels should indicate the most likely flow paths. The results of the initial office and field investigations should provide sufficient information to direct the final analysis.

SUMMARY

The previous three stages demonstrate that flood risk on alluvial fans is not unpredictable, but rather that it is predictable with varying degrees of uncertainty. The assumption of a uniform risk (FEMA, 1995) or complete uncertainty across an alluvial fan can be used as a formalized guess that allows one to delineate risk on the FIRM using a straightforward technique. This technique may be reasonable for the delineation of hazards on certain alluvial fans. The method proposed by Dawdy (1979) is an insightful application of the total probability equation. Although the assumptions used to solve the equation may vary for each situation, the method itself is sound and quite general.

A FIRM showing alluvial fan flooding hazards mapped considering complete uncertainty is of little use for floodplain management. By making a conservative trade-off in favor of all possibilities, this type of FIRM ignores the importance and the more threatening hazard of flow in existing channels and historical flow paths and conversely penalizes safer areas..

The FEMA *Guidelines and Specification for Study Contractors* (1995) asserts that flow paths for alluvial fan flooding are unpredictable and the assumption of uniform uncertainty must be used in the hazard delineation unless written approval is sought. Approaching the wide range of alluvial fan flooding conditions from the inflexible perspective of this special case is part of the reason for the conflict surrounding this matter.

The committee recommends that all efforts at mapping start with the existing channel. For situations where there is an entrenched channel on an alluvial fan, the uncertainty may be set aside. However, elsewhere the uncertainty associated with flow path direction might cause one to select FEMA's uniform risk method. For the majority of the cases, however, consideration of specific, foreseeable scenarios based on stages 1 and 2 make the most sense.

For some undissected fans, the assumption of uniform flow path uncertainty may apply. Such cases are not in the majority, and yet they are the only cases where the computer program FAN (FEMA, 1990) might be applicable.

REFERENCES

Bull, W. B. 1963. Alluvial-fan deposits in western Fresno County, California. Journal of Geology 71:243-351.

Bull, W. B. 1964. Alluvial fans and near surface subsidence in western Fresno County, California. U.S. Geological Survey Professional Paper 437-A. Reston, Va.: U.S. Geological Survey.

Bull, W. B. 1977. The alluvial fan environment. Progress in Physical Geography 1(2):222-270.

Bull, W. B. 1991. Geomorphic Response to Climatic Change. New York: Oxford University Press.

Christenson, G. E., and C. Purcell. 1985. Correlation and age of Quaternary alluvial-fan sequences, Basin and Range province, southwestern United States. Pp. 115-122 in Soils and Quaternary Geology of the Southwestern United States. GSA Special Paper 203. Boulder, Colo.: The Geological Society of America.

Cooke, R., A. Warren, and A. Goudie. 1993. Desert Geomorphology. London, England: University College London Press.

Dawdy, D. R. 1979. Flood frequency estimates on alluvial fans. American Society of Civil Engineers Journal of the Hydraulics Division 105(HY11):407-1413.

Dorn, R. I. 1994. The role of climatic change in alluvial fan development. Pp. 593-615 in Geomorphology of Desert Environments, A. D. Abrahams and A. J. Parsons, eds. London, England: Chapman and Hall.

Federal Emergency Management Agency (FEMA). 1990. FAN: An Alluvial Fan Flooding Computer Program, User's Manual and Program Disk. Washington, D.C.: FEMA.

Federal Emergency Management Agency (FEMA). 1995. Guidelines and specifications for study contractors. Document no. 37, Appendix 5: Studies of alluvial fan flooding, Washington, D.C.: FEMA.

Hereford, R., K. S. Thompson, K. J. Burke, and H. C. Fairley. 1995. Late Holocene debris fans and alluvial chronology of the Colorado River, Eastern Grand Canyon, Arizona. U.S. Geological Survey Open-File Report 95-97. Reston, Va.: U.S. Geological Survey.

Hydrologic Engineering Center (HEC). 1990. HEC-2 Water Surface Profiles, User's Manual. Davis, Calif.: U.S. Army Corps of Engineers Water Resources Support Center.

Keaton, J. R. 1988. A Probabilistic Model for Hazards-Related Sedimentation Processes on Alluvial Fans in Davis County. Ph.D. dissertation. Texas A&M University, College Station.

Kellerhals, R., and M. Church. 1990. Hazard management on fans, with examples from British Columbia. In Alluvial Fans: A Field Approach, A. H. Rachocki and M. Church, eds. New York: John Wiley & Sons.

MacArthur, R. C. 1983. Evaluation of the effects of fire on sediment delivery rates in a southern California watershed. In Proceedings of the D. B. Simons Symposium on Erosion and Sedimentation, Colorado State University, Fort Collins. July 27-29, 1983.

National Research Council. 1995. Flood Risk Management and the American River Basin: An Evaluation. Washington, D.C.: National Academy Press.

Pearthree, P. A., K. A. Demsey, J. Onken, K. R. Vincent, and P. K. House. 1992. Geomorphic Assessment of Flood-Prone Areas on the Southern Piedmont of the Tortolita Mountains, Pima County, Arizona. Arizona Geological Survey Open-File Report 91-11. Tucson, Ariz.: Arizona Geological Survey.

Ritter, J. B., J. R. Miller, Y. Enzel, S. D. Howes, G. Nadon, M. D. Grubb, K. A. Hoover, T. Olsen, S. L. Reneau, D. Sack, C. L. Summa, I. Taylor, K. C. N. Touysinhthiphonexay, E. G. Yodis, N. P. Schneider, D. F. Ritter, and S. G. Wells. 1993. Quaternary evolution of Cedar Creek alluvial fan, Montana. Geomorphology 8:287-304.

Ritter, D. F., R. C. Kochel, and J. Miller. 1995. Process Geomorphology, 3rd Ed. Dubuque, Iowa: Times Mirror Higher Education Group.

Skinner, B. J., and S. C. Porter. 1995. The Blue Planet. New York: John Wiley & Sons.

U.S. Army Corps of Engineers (USACE). 1992. Guidelines for Risk and Uncertainty Analysis in Water Resources Planning. Report 92-R-1. Fort Belvoir, Va.: USACE Water Resources Support.

Wells, S. G., and A. M. Harvey. 1987. Sedimentologic and geomorphic variations in storm-generated alluvial fans, Howgill Fells, northwest England. The Geological Society of America Bulletin 98:182-198.

4

Applying the Indicators to Example Fans

Not all alluvial fans, or those geologic features that are commonly believed to be alluvial fans, are subject to alluvial fan flooding. To show how dramatically such sites can vary, the committee selected seven sites for in-depth analysis and applied the indicators presented in Chapter 3. The sites represent a wide range of flood processes, from unconfined water flooding and debris flows on untrenched active fans to confined water flooding in fully trenched inactive alluvial fans. Six alluvial fans in the western United States are used to illustrate different flood processes, and a group of fans in Virginia illustrate a particular type of flood hazard in the eastern United States (Figure 4-1). By applying the indicators to each of the example sites, the committee was able to see whether or not the site meets the criteria suggested earlier in the proposed definition of an alluvial fan. In addition, insights are gained about how the definition and the indicators function in the field, and the advantages and disadvantages to those who ultimately will have to apply the guidance in a regulatory context.

Each of the examples also represents a different amount of study (Table 4-1). The Arizona examples show the problems faced in major urbanizing areas where there is intense interest and resources to support detailed investigation. The California examples represent a modest amount of study that included a brief field reconnaissance of each fan and compilation of geologic, topographic, and soil maps and aerial photographs. A similar approach was used to characterize the Utah fan, including examination of many technical reports produced following unusual flooding of 1983 and 1984. An exhaustive study of technical literature was the basis to typify the Virginia fans, which illustrate that alluvial fan flooding is not strictly a western phenomenon. Four of the sites were inspected by the committee, and the other three sites were inspected by at least one committee member. However, the committee wants to emphasize that it did not conduct a thorough field investigation of any site, as would be required for regulatory purposes, and thus these examples are purely illustrative and not intended to influence decision making on these fans.

HENDERSON CANYON, CALIFORNIA

The Henderson Canyon alluvial fan, which is located in eastern San Diego County near Borrego Springs, California, is below a drainage basin of approximately 16.6 km^2 (6.40 mi^2) that

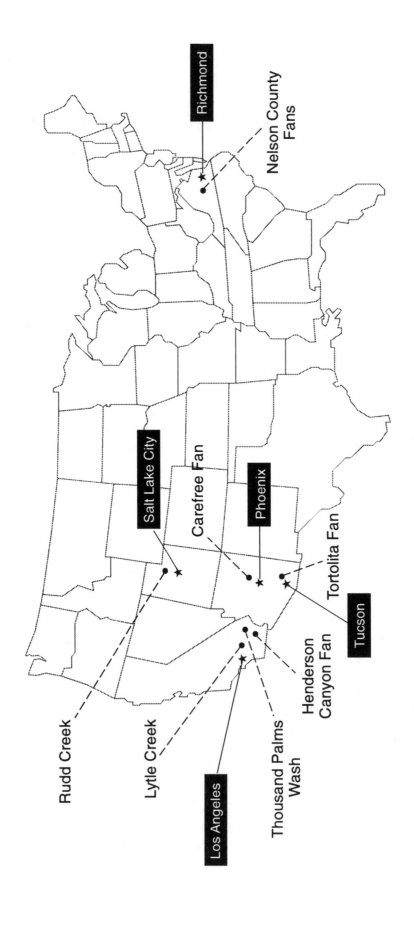

FIGURE 4-1 Six fans in the western United States are used to illustrate different flood processes, and a group of fans in Virginia illustrate a particular type of flood hazard in the eastern United States.

Table 4-1 Amount of Study, Sedimentation, Major Flood Processes, Flow Path Movement, and Relevant Comments for Example Fans (sites are listed in order from the West Coast of the United States)

Site	Amount of Study	Sedimentation	Major Process	Subject to Alluvial Fan Flooding	Comments
Henderson Canyon, California	Modest	Inactive and active	Water flood	Yes	Flooding confined to large trenches on relict fan. Sheet flooding on active fan. The use of maps, aerial photographs, soil surveys, and field reconnaissance is described in the example. See Appendix A.
Thousand Palms, California	Modest	Active	Water flood	Yes	Sheetflooding on fan. The general alignment of the fan has been altered by faulting.
Lytle Creek, California	Modest	Inactive	Water flood	No	Flooding confined to a single large trenched channel. See Appendix A.
Tortolita Mountains, Arizona	Extensive	Inactive	Water flood	No	Network of flow paths has appearance of active fan, but flow paths were stable during major flood. See Wild Burro alluvial fan in Appendix A.
Carefree, Arizona	Extensive	Inactive	Water flood	No	Flow is confined to network of trenched distributary channels with no evidence of flow path movement. The use of soil surveys by the Natural Resources Conservation Service is described in the example. See Appendix A.
Rudd Creek, Utah	Average	Active	Debris flow and water flood	Yes	Major debris flow in 1983 damaged or destroyed many homes. Episodes of debris flows are on the order of once every 100 to 1,400 years. See Wasatch Front alluvial fans in Appendix A.
Nelson County, Virginia	Average	Active	Debris flow and water flood	Yes?	Episodes of debris flows are on the order of once every 3,000 to 4,000 years.

heads on the eastern slopes of rugged mountains at an elevation of 1,420 m (4,659 feet) (Figures 4-2 and 4-3). Most of the mountainous basin above an elevation of 488 m (1,601 feet) is practically barren of vegetation, and runoff from the steep slopes is rapid. The region is tectonically active, but active faulting is generally located to the north and south of the basin and alluvial fan (Sharp, 1972). The alluvial fan is an example of an arid-clime composite fan with both relict debris flow and modern water flow processes where hazards on the relict fan have been significantly altered by geologically recent channel trenching.

Recognizing and Characterizing Alluvial Fans

Determining Whether or Not a Landform Is an Alluvial Fan

This landform, known as the Henderson Canyon alluvial fan, was identified as an alluvial fan using the criteria defined in Chapter 3 for material composition, morphology, and location.

Composition   The site is identified as "sloping gullied land" of an alluvial fan on National Resource Conservation Service (NRCS, 1973) soil maps on a 7.5-minute series orthophoto base. It consists of alluvial sediments derived from igneous, sedimentary, and metamorphic rocks. Carrizo soil, which is gravelly sand derived from granitic alluvium and is also associated with alluvial fans, is shown on the lower fan near the valley. The type and relative position of the mapped soils suggest entrenched channels in a relict fan with an active fan downslope in the Carrizo soil.

Only upon field inspection of the alluvial fan was it clear that modern channels are deeply trenched into relict debris deposits of the sloping gullied land shown on the soil map (NRCS, 1973). Numerous massive mounds of debris flow deposits are composed of many 0.15- to 0.9-m (0.5 to 3.0-foot) boulders, and many of the fines have been washed from the debris matrix, forming sandy interlobe areas (Figure 4-4). Deposits are massive with distinct boundaries readily observed by field inspection; there is inverse grading and there is a concentration of large boulders at the snout of the deposited lobes.

Two major entrenched channels combine in the relict material to form a single entrenched channel that leads to modern alluvial deposits of gravelly sand and scattered cobbles. On the modern deposits, there is some stratification of thin beds that appear to have been deposited as large sheets. These loose and friable deposits are mapped as Carrizo type soil. The fan is thus composed of relict and modern alluvial deposits like those of an alluvial fan.

Morphology   The site has the general appearance of a sector of a cone with concentric contour lines that are generally convex downslope and laterally confined in an "embayment" within the general alignment of a mountain front. The form and general bounds of the site can be readily identified on the available 7.5-minute series USGS topographic map (Borrego Palm Canyon Quadrangle). The landform is shaped like a partly extended fan that attenuates at the lateral bounds of a large valley to the east.

Location   The Henderson Canyon fan is located at a topographic break in lateral confinement at the upper end of the embayment. Many such breaks, although subtle, were observed by

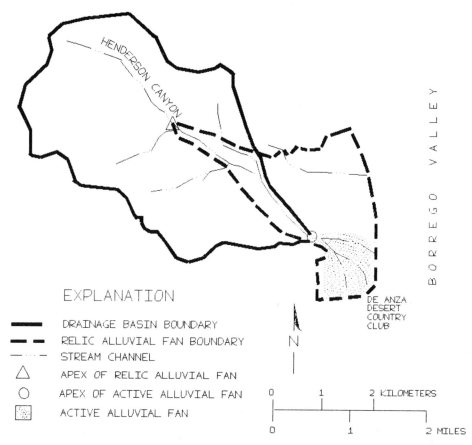

FIGURE 4-2 Henderson Canyon drainage basin showing relict alluvial fan boundaries, location of active alluvial fan, location of apex of relict fan, and location of apex of active fan.

Hjalmarson and Kemna (1991) using channel profiles of the change in channel slope between topographic map contours. Hooke (1967) described this flattening and steepening of channel slope where confinement is lost at the apex (or intersection point). This break is not apparent on the channel profile defined using the USGS topographic map (Figure 4-3) possibly because the profile represents modern drainage effects. Rather, the surface of this relict fan is a few meters above the present trenched stream channel as observed on the USGS topographic map when used in conjunction with color-infrared aerial photographs obtained from the EROS Data Center of the USGS and orthophoto maps. The topographic break is located near the confluence of two major mountain streams at the upslope edge of relict alluvial deposits. There is a significant change in the surface texture at this location as shown on the aerial photographs and orthophoto maps.

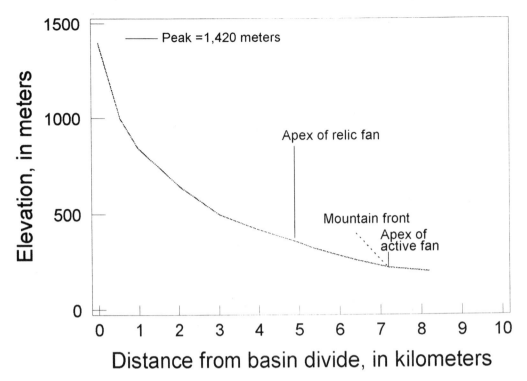

FIGURE 4-3 Profile of Henderson Canyon relict alluvial fan near Borrego Springs, California.

FIGURE 4-4 View looking downslope and across to the south from center of Henderson Canyon relict alluvial fan at typical boulder mound 1.2 to 1.8 m (3.9 to 5.9 feet) high, March 21, 1995. Courtesy of H. W. Hjalmarson.

Defining the Lateral Boundaries and Topographic Apex of the Alluvial Fan

The lateral bounds are at the toe of steep-bedrock mountain slopes that form the embayment. Below the mountain front the lateral bounds are defined by topographic ridges between fan drainage channels and the adjacent drainage channels. A swale-like drainage that traverses the fan at the western edge of Borrego Valley forms the fan toe. The fan toe closely coincides with the lower limits of the Carrizo soil unit shown on the NRCS soil survey maps. The general bounds of the alluvial fan can be readily identified on the available 7.5-minute series USGS topographic map (Borrego Palm Canyon Quadrangle).

Defining the Nature of the Alluvial Fan Environment

A few large channels and an area of active sedimentation are readily apparent on the color-infrared aerial photographs. The location of these channels coincided with evidence of trenched channels in the upper- and mid-fan areas, as suggested by the saw-toothed appearance of the contour lines on the USGS topographic map. The few large kinks in the contours that point upslope are typical of a fan surface with large incised channels. The field investigation revealed light grayish colored rock on the bed and banks of trenched channels, which is indicative of recent abrasion during sediment transport. The adjacent boulders in the debris lobes were lightly covered with rock varnish, which is indicative of a stable surface.

Field examination was needed to precisely locate the wide, flat hydrographic apex of the active alluvial fan located on the right, or south, side of the relict fan (Figure 4-2). The hydrographic apex is located at a gradual hydraulic expansion at the end of the large trenched channel.

Active Fan

There is active sedimentation below the hydrographic apex and active erosion above it. The average slope of the active fan is about 0.025 over a length of about 1.2 km (0.7 mi). The poorly defined channels are slightly braided, with large width-to-depth ratios in the upper fan. There are large sheetflood areas in the middle fan. The surface material appears to be very recent and has not developed a soil. The width of the active fan is approximately one-third the total width along the toe of the Henderson Canyon alluvial fan.

The lateral boundaries of the active fan were apparent on orthophoto and topographic maps at 1:2400 scale with 5-foot (1.52 m) contour intervals (furnished by the San Diego County Department of Public Works). These maps were available for the area below the major incised channels that included most of the active fan. The boundaries of the active fan area could be defined on these maps, especially when used in conjunction with the color-infrared aerial photographs. A few relict debris lobes also were indicated on the large-scale topographic maps, but most were attenuated and indistinct. Many sheetflood paths are clearly shown on the photographs, but the lateral bounds of sheetflood are indistinct. A field inspection was needed to distinguish between the sheetflood deposits of the active fan and the deposits below the attenuated debris lobes of the relict fan. Both of these deposits are mapped as Carrizo-type soil (NRCS, 1973).

The upper (hydrographic apex) and lower limits (fan toe) of the active fan are approximately defined by the mapped limits of the Carrizo-type soil. However, based on a field inspection of recent sheetflood deposits, much of the toe is below the Carrizo soil at or slightly upslope of a transverse swale-like drainage at the western edge of Borrego Valley. The toe is indistinct, partly because the urban land has been altered by construction or obscured by structures, pavement, and golf courses.

Relict Fan

The mid-fan is composed of limey debris flow deposits that have been exposed along the entrenched channels. The exposed carbonates of the B horizon extend to depths of nearly 2 m (7.6 feet) along the steep banks of the entrenched channels and are indicative of soils that are more than 10,000 years old in this arid region (Machette, 1985).

There is no channel formation or other evidence of flow along the downfan side of the entrenched channel. Evidence of channel formation between the debris lobes is apparent only several hundred meters down the relict fan from the transverse channel. The few small channels in this area are formed by local runoff. The attenuated debris lobes in this area become undefined in the vicinity of the Carrizo soil downslope.

Mountainous Drainage Basin

Field observations of debris sources on the mountainsides revealed that most of the upper slopes are bare rock and apparently too steep for debris accumulation. The soil in the mountains is loamy coarse sand in texture and is sparse and shallow. The lower slopes of the mountains are covered with boulder debris that appears to be stable because the rock is covered with dark desert varnish and the slopes are less than the angle of repose of the rocks. No slumping was observed. There is one site of a geologically recent small debris slide on the southern side of the drainage basin where the hillslope is at least 38 degrees. This slide is apparent because the scar appears geologically fresh among the darkly varnished surrounding bedrock. Deposited rock from this slide is far from active channels and is not a significant source of material for debris flows down the fan. Little, if any, debris along the stream channels can be seen on aerial photographs. Because there is no known history of recent debris flows in the area and there is little evidence of geologically recent debris flow potential, large debris flows of the size that produced the many mounds are considered unlikely.

Storm of August 15-17, 1977

Tropical Storm Doreen produced from 7.6 to 12.7 cm (3.0 to 5 in) of rainfall in the vicinity of the Henderson Canyon alluvial fan. Most of the rain fell during a few hours on the evening of August 16, 1977, producing a peak discharge of 90.6 m^3/s (3,200 feet^3/s) at the apex of the active alluvial fan about 0.8 km (0.5 mi) upslope of the De Anza Desert Country Club (San Diego County, 1977). This peak discharge was nearly equal to the 100-year flood based on methods described by Thomas et al. (1994). The flow split into two distinct paths upstream of the community. A short distance below the hydrographic apex, where the flow split the floodwaters became less confined and apparently coalesced as sheetflooding. The distribution of floodwater across the active fan at any particular time is unknown, but nearly all of the active fan was inundated at one time or another during the flood. Aerial photographs of flood remnants and hydraulic computations of peak discharge amounts suggest that floodwater covered most of the active fan at the time of the peak discharge. Large amounts of sand with gravel and a few small boulders were deposited throughout the community, and some floodflow passed through the Country Club, inundating farmland to the east. About 100 homes were damaged as previously effective drainage ditches and debris dams were overwhelmed by the floodwater and debris.

Floodwater from the drainage basin was conveyed in the incised channels on the relict fan to the hydrographic apex. Most of the basin flow was in the center channel and all downfan flow from the drainage basin was intersected by the channel, crossing the fan from the north.

Changes in Flow Path

A comparison of entrenched stream channels depicted on aerial photographs showed no discernible channel movement, enlargement, or formation on the relict fan. Three sets of aerial photographs obtained from the EROS Data Center of the USGS were used for the comparison. The photographs were good-quality black and white for 1954, poor-quality color-infrared for 1971, and excellent quality color-infrared for 1990. The large-scale orthophoto maps mentioned previously also were used for the assessment of changes in flow path.

Significant flow path change was not apparent on the active fan, but minor change of the sheetflood paths is suggested on the photographs and orthophoto maps. Because the paths are obscured by vegetation and possibly by eolian effects, the amount of movement is uncertain.

Characterizing Alluvial Fan Flooding Processes

Floods have eroded and apparently will continue to erode relict fan material and deposit it on the active fan. All the evidence points to the conclusion that the De Anza Desert Country Club, which is located on the lower portion of the active fan, is in the direct path of future sediment-laden water floods emanating from the 16.6-km^2 (6.4-mi^2) basin (Figure 4-2). The flood risk on the active fan is much greater than would be predicted by application of the FEMA procedure without recognition of the distinction between the active and the relict portions of the fan.

Defining Areas of Active Alluvial Fan Flood Hazard

Floodwater leaves the confines of the trenched channel at the hydrologic apex and spreads in two swales as sheetflood. The entire area of the active alluvial fan (Figure 4-2) is subject to alluvial fan flooding.

Where does flow depart from confined channels? Below the hydrologic apex the flow paths are difficult to predict because flow is shallow and unconfined.

Where does sheet flow deposition occur? Sheet flow deposition of sediment over most of the active fan, like that for the flood of August 16, 1977, can be expected during a single flood. Small amounts of sediment deposition on the upper active fan may cause changes in the paths of flow because the flood depths are small and flow is unconfined.

Where does debris flow deposition occur? There is no evidence of recent debris flows.

Where are there structures or obstructions that might aggravate or cause alluvial fan flooding? Nearly all of the area to the west of the hydrologic apex is within the Anza-Borrego Desert State Park and is not subject to development. The urban development of the De Anza Desert Country Club located on the lower portion of the active fan may alter flow paths and concentrate floodflow in streets and other open areas.

Where can the flood hazard be mitigated by means other than major structural flood control measures? Only major structural controls will be effective because development is along the entire lower portion of the active fan.

Defining Areas of Nonalluvial Fan Flooding Hazard Along Stable Channels

Much of the surface of the relict fan is above the level of flooding in the trenched channels and is not subject to alluvial fan flooding. Floodflow from the surrounding mountains is confined to trenched channels that traverse the relict fan. Fundamental hydraulic computations of channel capacity confirm this conclusion. Crude estimates of channel roughness, size, and slope using a hand level and surveying rod show that the channel capacity of the major channels is several times that needed to convey the 100-year flood, estimated using methods by Thomas et al. (1994), across the surface to the apex of the active fan.

Most of the runoff below the transverse channel has been from the relict fan itself, with a small amount from a small mountain basin to the north. Floodflow is less confined downfan in this area where the debris mounds are small and intermound areas have filled with sediment. Some sheetflooding and sedimentation along the toe of the relict fan are expected.

Determining the Type of Processes Occurring on the Active Parts of the Alluvial Fan

A brief field inspection of the deposited material of the active fan suggested the action of water flood processes in the following ways: Sheetflooding is indicated in the mid-fan area by the thin sorted beds of sand with silt and some gravel, which were loose and friable, and appeared continuous over large areas. The channels in the upper fan had very large width-to-depth ratios, indicating water flow. The deposits were permeable. There are no massive and unstratified deposits and no channels or debris mounds in the middle and lower fan that indicated debris flows. No indicators of debris flows (see Table 3-3 and Figure 3-8) were observed on the active fan.

THOUSAND PALMS WASH, CALIFORNIA

Thousand Palms fan is located in Riverside County, California. Runoff from the Little San Bernardino Mountains to the northeast is collected along the Indio Hills, mostly by Deception Wash, a tributary to Thousand Palms wash, at the Mission Creek fault. The drainage area for Thousand Palms wash is 217 km^2 (84 mi^2). As the wash passes through the Indio Hills and crosses the San Andreas fault zone into the Coachella Valley, it flows onto a broad alluvial fan (Figure 4-5). There is a general lack of soil development and vegetation on the fan. Furthermore, the main channel loses definition shortly after passing through the apex.

Recognizing and Characterizing Alluvial Fans

The landform is identified on NRCS (1980) soil maps as an alluvial fan. The revised definition was applied to this example, and the landform was found to be an alluvial fan.

Determining Whether or Not a Landform Is an Alluvial Fan

The fan shape is apparent from topographic maps. Some of the soil material is from upstream alluvial fan deposits that have been removed by headcutting in response to strike-slip movement at the Mission Creek fault. The loose and friable deposits are in sheets or beds of sand and silt. The fan has classic concentric contours, but the center of the fan bends gradually to the left or east. Thus, the upper part of the fan faces southwest, while the lower part faces more to the south and east. This shape is related to the general morphology of the Coachella Valley, which lies to the south, and probably to slip faulting across the middle of the fan. There is considerable channel widening as Thousand Palms wash leaves the confines of the Indio Hills and crosses the San Andreas fault, where there has been vertical and lateral displacement.

The landform has the composition, morphology, and location to meet the committee's criteria for an alluvial fan.

Defining the Lateral Boundaries and Topographic Apex of the Alluvial Fan

The lateral bounds of the upper part of the fan are at the toes of steep slopes of older alluvial deposits that can be readily identified on the available 7.5-minute series USGS topographic map. Beyond the general alignment of the Indio Hills, the lateral bounds are the topographic trough lines on each side of the fan. These boundaries are swales and appear slightly concave down-fan on the topographic map. The bounds generally correspond to the Carsitas and Myoma soils that are associated with alluvial fans (NRCS, 1980a). The western boundary is indistinct in places because of wind-blown deposits of sand and silt. The fan coalesces with a small fan to the west and with two small fans to the east.

FIGURE 4-5 Aerial photograph of Thousand Palms wash (1993). Courtesy of Aerial Fotobank, Inc.

Defining the Nature of the Alluvial Fan Environment

Active Fan

The topographic apex is located at the highest point on the fan where floodflow leaves the confines of a wide sand channel. The slope of the fan is approximately 2.5 percent near the apex and 0.5 percent at the margin where there is a transition to wind-blown sand dunes. The median sediment grain size near the apex is approximately 1 mm (0.03 in) and, at the margin of the fan, 0.2 mm (0.007 in). No soil development was observed during a brief field inspection of the fan.

Relict Fan

There is no visible relict fan.

Mountainous Drainage Basin

The upper part of this complex basin is in the Little San Bernardino Mountains to the north. Three large alluvial fans have formed along the southern slopes of the mountains within the basin. Because of movement along the Mission Creek fault, the toes of the fans are gullied. Remobilized sediment from these fans is a major source of sediment for the Thousand Palms alluvial fan.

Flood of 1977

An aerial photograph of the fan taken shortly after the flood of 1977 depicts paths of recent sheetflooding over much of the upper part of the fan. It is unknown if there was simultaneous inundation over the fan as suggested by the flow lines. About 1 mile below the apex to the southeast, the flow paths were separated by small islands of sand dunes. The dune islands become larger downfan. About 2 to 3 miles below the apex the flow becomes channelized. Because wind-blown sands obscure flow patterns following large floods, the widespread flooding depicted on this aerial photograph, on file at the Coachella Valley Water District, is a primary basis for concluding the fan is active and subject to sheetfloods.

Flow Path Changes

Flow path movement is likely during major floods because of the low transverse relief, undeveloped soils, and evidence of channel bed aggradation where coarse deposits may force water overbank and form new paths.

Characterizing Alluvial Fan Flooding and Sedimentation Processes

The braided flow paths appear uncertain just downstream of the topographic apex. Evidence of sheetflood is apparent from both field inspection and review of historical accounts of flooding on file at the Coachella Valley Water District. Water flood processes are suggested by the stratified deposits of silt, sand, and some gravel which are highly permeable and friable. There are also scattered cobbles in the upper fan. The channels in the upper and middle fan have very large width-to-depth ratios. There are no debris mounds or unstratified cemented deposits that are characteristic of debris flows.

There are defined flow paths on the alluvial fan that convey floods of smaller-magnitude. The absence of vegetation and lack of soil development on the fan indicate that recent alluvial flood deposits are extensive. The area subject to sheetflooding closely corresponds to the Carsitas cobbly sand soil (NRCS, 1980a). Floods of large magnitude are subject to flow path uncertainty on the upper part of the fan.

The relatively steep fan slope and the absence of topographic confinement create a condition where measures such as set backs or elevation on fill may not reliably mitigate flood hazards on much of the fan. Structural mitigation of the flood hazard is required.

Thousand Palms Wash fits both the existing and the proposed definition for alluvial fan flooding.

LYTLE CREEK, CALIFORNIA

Lytle Creek is located in San Bernardino County, California. At the topographic apex, it drains approximately 50 square miles of the San Gabriel Mountains, which are composed of highly fractured rock at steep slopes. Erosion from the watershed produces a high yield of very coarse sediment. The fan slope is almost 3 percent. The main channel is incised as it leaves the mountains (Figure 4-6). Since the 1940s a series of spur dikes and levees have been built to confine the flows to a narrow corridor along the fan. Lytle Creek eventually combines with Cajon Creek before entering the Santa Ana River.

This fan is one of a series that consist of unconsolidated alluvial deposits to the south of the Sierra Madre fault zone. The piedmont is a complex bajada with several trenched and untrenched alluvial fans. For example, the Cucamonga fan, to the west, is deeply trenched, but the nearby Day and Deer creek fans, also to the west of this site, are not entrenched and have areas of active sedimentation and flooding.

Recognizing and Characterizing Alluvial Fans

This landform was identified as an alluvial fan using the committee's criteria (defined in Chapter 3) for material composition, morphology, and location. This landform was identified by both Eckis (1928) and the Natural Resources Conservation Service (1980b) as an alluvial fan.

Determining Whether or Not a Landform Is an Alluvial Fan

The fan is composed of gravelly and bouldery granitic alluvium and other sediment including limestone, schist, and volcanic fragments of the San Gabriel Mountains to the north (Eckis, 1928). Eckis mapped the feature as dissected recent alluvium of an alluvial fan. The soil is composed of stony loamy sand deposited on alluvial fans (NRCS, 1980b).

The landform has the appearance of a cone nearly fully extended. The concentric contours are convex downslope, with some lateral restriction on the left or east side. The form and general

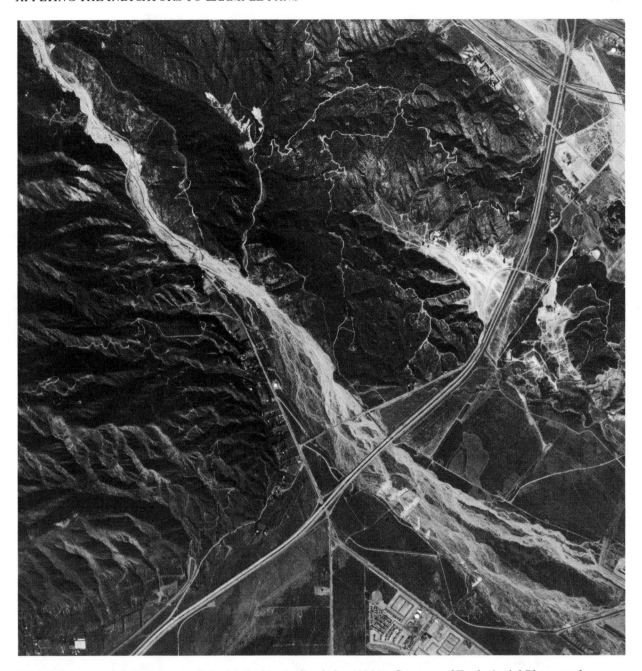

FIGURE 4-6 Aerial photograph of the head of Lytle Creek fan (1991). Courtesy of Eagle Aerial Photography.

bounds can be readily identified on the USGS 7.5-minute series topographic map (Devore Quadrangle).

   The apex of this fan is at the southern front of the San Gabriel Mountains near the eastern edge of the Cucamonga scarp, which is associated with a few thousand feet of differential movement. A hint of differential vertical movement is suggested by the steepening of the elevation profile as the stream emerges from the confines of the V-shaped canyon upstream of Interstate

Highway 15. Much of the topographic break, however, also is associated with lateral unconfinement as the stream leaves the confines of the steep mountain canyon onto the piedmont.

Defining the Lateral Boundaries and Topographic Apex of the Alluvial Fan

The fan is broad and long, and the eastern lateral bounds are at the mountain in the upper part of the fan and at Cajon wash in the middle and lower parts. The western boundary is at a swale where the fan coalesces with the San Sevaine Canyon, Etiwanda Creek, and Day Creek fans. These bounds are roughly defined on the NRCS (1980b) soil survey maps, which identify several soils as those of an alluvial fan. The topographic apex is about 305 m (1,000 feet) within the mountain where the contours change from concave to convex. The apex is about 1830 m (6,000 feet) downstream of the U.S. Geological Survey streamflow gage.

Defining the Nature of the Alluvial Fan Environment

Active Fan

There is no active fan.

Relict Fan

The relict fan is being dissected by a modern fanhead trench that follows a regional fault. Differential vertical movement of this fault is suggested by different bank heights near the Santa Ana River a few miles downstream (Eckis, 1928). Because of this 2.5- to 5-m-deep (8- to 15-foot-deep) by about 610-m-wide (2,000-foot-wide) trench that dissects the relict fan from the mountains to Cajon Wash, no areas of this fan are subject to active sedimentation. The capacity of the trench is many times that needed to convey the 100-year discharge.

Mountainous Drainage Basin

A cursory examination of sediment accumulation and likelihood of slope failure suggests that there may be potential for a large sedimentation event such as a debris flow. For example, there is a remnant of a large slope failure to the west in the Day Creek basin, where there is little channel trenching and much of the fan is active. It has been a considerable time since a major sedimentation event, and it is possible that basin conditions are evolving toward such an event. A large event might be triggered by a large wildfire followed by heavy precipitation. A comprehensive assessment is beyond the scope of this example. However, tentatively, a large sedimentation event is considered so infrequent to be outside the realm of traditional hazard consideration.

Past Floods

Historical flooding in Lytle Creek between the apex of the old fan at the front of the San Gabriel Range and the Santa Ana River to the south has been confined to the incised channel (also see Appendix A). The channel has cut into unconsolidated alluvial deposits from the canyon mouth near the apex of the old fan to the El Cajon Wash and even further downstream to near the confluence with the Santa Ana River. Some lateral movement of the steep cut banks is suggested by reports of bridge failures during past flooding at San Bernardino (McGlashan and Ebert, 1918) and reports of channel movement just above the canyon mouth (Troxell, 1942). The largest known discharge from the 119.9 km^2 (46.3 mi^2) drainage basin above the U.S. Geological Survey stream gaging station near Fontana, California (Number 11062000), was 1,017 m^3 (35,900 feet^3) per second on January 25, 1969 (Chin et al., 1991). The capacity of the channel about 2.4 km (1.5 mi) below the gaging station and just below the mountain front is about 3 times the magnitude of the 1969 flood.

Flow Path Changes

The present incised channel (1995) of Lytle Creek is very similar to the channel reported by Eckis (1928). DMA Consulting Engineers (1985) also reported that the flow path of major floods in upper Lytle Creek was unchanged from 1935 to 1969. A comparison of flow paths shown on aerial photographs of October 12, 1967 (DMA Consulting Engineers, 1985), and an aerial photograph of August 29, 1989, indicates no movement of the channel banks. Thus, using channel conditions suggested in the account of flooding by Troxell (1942), the channel of Lytle Creek has been deeply incised since at least 1862, and the path of flow has not changed.

The soil has a well-developed, grayish-brown surface layer of stony or gravelly loamy sand. The underlying material is brown, very stony sand to a depth of 1.5 m (5 feet). There is no active surface on the Lytle Creek alluvial fan.

Characterizing Alluvial Fan Flooding Processes

The Lytle Creek alluvial fan is not subject to alluvial fan flooding processes as defined by the committee. The incised channel conveys flows at shallow depths and high velocity during floods. Flood control structures installed over the past 40 years have prevented large-scale channel migration. Although substantial erosion and deposition occur in the main channel, flow path uncertainty is minimal.

The floodplain is topographically bounded on both sides. This confinement is mostly because of channel incision and to a lesser degree because of the spur dikes and levees a few hundred feet above and below Interstate Highway 15. Without the presence of the constructed spur dikes and levees, flood mitigation measures such as setbacks and elevation on fill might not reliably mitigate the hazard.

Lytle Creek is entrenched in a classically shaped relict alluvial fan. It does not, however, fit the committee's definition of alluvial fan flooding. Sediment delivered by watershed erosion processes is conveyed through the fan, but large sedimentation events are tentatively considered remotely possible. The surface of the relict fan is resistant to erosion, and parts of the surface may be flooded by local runoff. This does not, by itself, mean that this area is subject to alluvial fan flooding, however.

TORTOLITA MOUNTAINS, ARIZONA

The Tortolita Mountain piedmont lies 20 km (12.4 mi) northwest of downtown Tucson, Arizona. The region is semiarid, with 280-mm (1.1-in) average annual precipitation. Average annual temperature is 20.3 degrees Celsius. Summer monsoonal precipitation results in intense floods. The area is characterized by sloping alluvial surfaces extending approximately 10 km (6.2 mi) from the front of the Tortolita Mountains to the floodplain of the Santa Cruz River (Figure 4-7). The highest elevation in the Tortolita Mountains is 1,533 m (5,030 feet). Elevation at its mountain front is approximately 900 m (2,953 feet), and the Santa Cruz valley floor elevation is approximately 600 m (197 feet). The Tortolita Mountain front was originally a steep fault scarp, but this range has been tectonically inactive since the late Tertiary. There is no modern fault scarp, and the mountain front is highly sinuous. Because this mountain front is inactive, deposition on the piedmont during the Quaternary has been controlled primarily by climate change (Fuller, 1990).

Recognizing and Characterizing Alluvial Fans

Determining Whether or Not a Landform Is an Alluvial Fan

The Tortolita piedmont consists mainly of the dissected remnants of ancient fans. These do not meet the committee's definition of an alluvial fan. However, there are subunits of the piedmont that do display the criteria. An example is Cottonwood fan (see box in Figure 4-7).

Composition The piedmont is mantled by alluvial sediments. The dissected fan remnants close to the Tortolita Mountain front are composed of gravel and boulders. The surfaces of these remnants are weathered to form desert pavements and soils. The latter include accumulations of fine material blown in by wind and the alteration products of chemical weathering over long periods of geologic time. These coarse-grained deposits decrease in grain size distally from the mountain front.

The alluvial fans that are inset into these dissected remnants are composed mainly of sand. This is because the fan apexes are not at the mountain front. The apexes are in the active stream channels that are incised through the old piedmont surfaces. The fine sediment is derived largely from the old soils that characterize the dissected fan remnants, which make up the divides between active stream channels cut into the piedmont.

Morphology The Tortolita piedmont consists of old alluvial surfaces that slope away from the mountain front. The surfaces were formed by ancient coalescing alluvial fans but are not considered fans today under the committee's definition because they lack the characteristic fan shape.

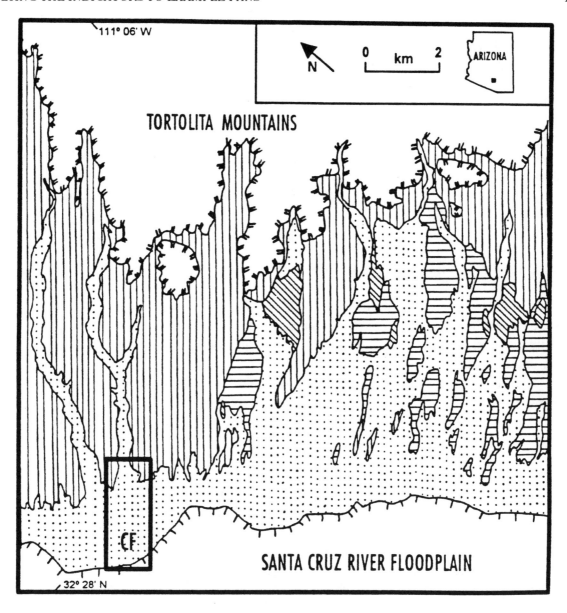

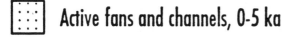 Active fans and channels, 0-5 ka

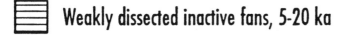 Weakly dissected inactive fans, 5-20 ka

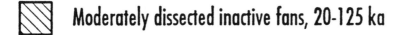 Moderately dissected inactive fans, 20-125 ka

 Highly dissected fan remnants, 125-750 ka

FIGURE 4-7 Geomorphologic map of alluvial surfaces on the southwestern piedmont of the Tortolita Mountains. CF indicates Cottonwood Fan. Dot pattern shows areas subject to active modern flooding (Unit b). The other units are old, inactive parts of the piedmont. SOURCE: Reprinted with permission from Field (1994).

Several active washes extend from major canyons in the Tortolita Mountains and are deeply incised into the old piedmont surfaces. As these washes extend beyond the ancient piedmont surfaces, they fan out radially. The pattern is obvious in active channelways visible on aerial photographs and recognized by historical documentation of active streamflow. An example is Cottonwood fan (Figure 4-8). This clearly shows a fan shape.

Location The Tortolita piedmont extends to the prominent topographic break of the Tortolita Mountains. However, this break is not associated with fan morphology but rather with the morphology of dissected old piedmont surfaces.

The Cottonwood fan extends headward to a confined channel that is incised into the old dissected piedmont surface.

Defining the Lateral Boundaries and Topographic Apex of the Alluvial Fan

The Tortolita piedmont does not have clearly defined lateral boundaries. The old piedmont surfaces extend back up the embayed elements of the mountain front (Figure 4-7). However, away from the mountain front these old dissected surfaces coalesce to form an apron sloping away from the mountains.

Cottonwood fan is bounded laterally by the depositional channels that head back to its upstream source. Lateral to these channels are either (1) old piedmont surfaces at higher elevations or (2) active channels that head back to other sources. The apex of the Cottonwood fan is clearly seen where the fan channels converge headward into a single thread (Figure 4-8).

Defining the Nature of the Alluvial Fan Environment

The extent of active channels on the Tortolita Mountain piedmont can be determined from geomorphologic study of aerial photographs, soil development, and topography (Figure 4-7). The active areas, frequently inundated by modern floods, have poorly developed soils, distributary drainage paths, and a lack of desert pavement or rock varnish. These areas are inset at lower relative elevations in relation to adjacent ancient piedmont surfaces. The latter show strong soil development, well-developed dendritic drainage (dissecting the old surfaces), closely packed desert pavement, and well-developed rock varnish. Detailed discussion of these distinguishing criteria is provided by Pearthree et al. (1992).

CAREFREE, ARIZONA

Carefree fan is located in central Arizona about 6.5 km (4.0 mi) south of the town of Carefree and a few kilometers north of Phoenix. Above the apex, the 11.1-km^2 (4.2-mi^2) long, narrow drainage basin heads on a relatively gently sloping pediment at 1,000-m (3,281-foot) elevation in desert vegetation. The apex is 14.5 km (9 mi) southwest of the head of the basin at an elevation of 649 m (2,129 feet). The area of the 7.1-km-long (4.4-mi-long) alluvial fan below the apex is 9.9 km^2 (3.8 mi^2) (Figure 4-9). Such pediments and alluvial deposits in central and

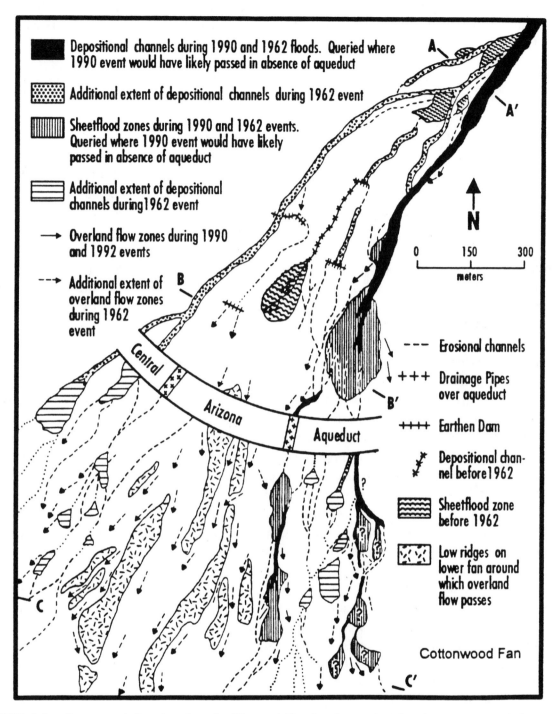

FIGURE 4-8 Cottonwood fan (see box in Figure 4-7), showing active fan features mapped from aerial photographs, historical accounts, and geomorphologic survey. SOURCE: Reprinted with permission from Field (1994).

southern Arizona are predominantly modern sediment transport surfaces. Relatively inactive alluvial fans are common on these geologically and hydraulically complex surfaces, where fan apexes typically are on the dissected piedmont near the lower edge of a pediment.

Recognizing and Characterizing Alluvial Fans

Determining Whether or Not a Landform Is an Alluvial Fan

This landform was identified as an alluvial fan using the committee's criteria (defined in Chapter 3) based on material composition, morphology and location. The crucial morphologic requirement was the planametric shape, because any transverse convex shape was lost to modern erosion.

Composition Most of the landform is alluvium, which has weathered to a developed gravelly loam and gravelly sandy loam on a fan terrace (Camp, 1986). Along the drainage ways are younger soils that are poorly developed sandy and gravelly loam (Anthony-Arizo and Antho-Carrizo-Maripo soil units, Figure 4-9).

Morphology The site has a general cone shape, but any transverse convex shape has been lost over much of the fan through erosion and coalescence with adjacent fans. Trenched distributary channels radiate downslope toward the west from the apex where the feeder channel leaves the confines of a gully. These channels are not readily apparent on USGS 7.5-minute topographic maps (Currys Corner and Cave Creek Quadrangles) or Natural Resources Conservation Service soil maps on a 7.5-minute series orthophoto base (Camp, 1986). The shape of the complex network of entrenched channels is clearly observed on large-scale topographic maps at 1:2,400 scale with 2 foot (0.61 m) contour intervals (furnished by the Flood Control District of Maricopa County, Arizona), especially when used in conjunction with color-infrared aerial photographs obtained from the EROS Data Center of the USGS. The landform is shaped like a partly extended fan that ends near a major stream channel. The stream channel appears to truncate the fan.

Location The landform is at the lower edge of a gradual transition zone between a pediment and *bajada* where floodflow leaves the confines of a gully and spreads laterally into two channels that are entrenched into old alluvium. A second small topographic break is located on the pediment about 4.8 km (3 mi) upstream where a small channel leaves the otherwise finite drainage basin bounds. This upstream diffluence is above the transition zone and is not considered part of this fan. Thus, the topographic break for this fan is downslope of all pediment remnants on the piedmont (Figure 4-9).

Defining the Lateral Boundaries and Topographic Apex of the Alluvial Fan

The topographic apex is located at the topographic break described above (Figure 4-9). The lateral boundaries of the fan were estimated using the large-scale topographic maps in conjunction with aerial photographs. The boundaries shown in Figure 4-9 generally are located at

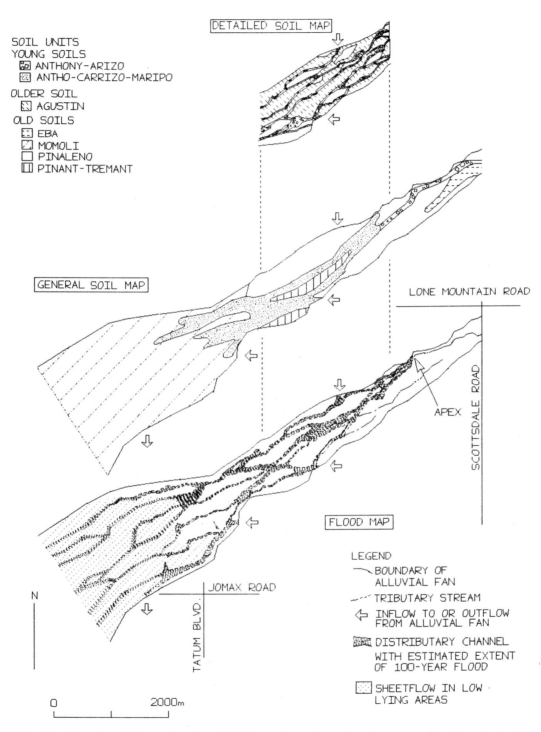

FIGURE 4-9 Carefree alluvial fan, showing distributary channels, soil units, and estimated extent of 100-year flood.

topographic ridges shared with adjacent fans. Mapping of the fan boundaries started at the apex and proceeded downslope. The lateral boundaries, which obviously are estimated downslope of any tributaries and distributaries, tend to be perpendicular to the general alignment of the topographic contours.

The fan boundary at the toe is indistinct and transitional between the dominantly distributary drainage pattern of the fan and the dominantly entrenched-tributary drainage pattern adjacent to Cave Creek. Below the fan toe the lower portion of the piedmont merges with Cave Creek to the west.

Defining the Nature of the Alluvial Fan Environment

The sediment balance appears to have turned negative, especially on the upper parts of the fan, where channel sedimentation has been replaced by erosion in the geologically recent past. In the lower parts of the fan the balance appears less negative (less net erosion), as suggested by two small areas of sheetflow and sediment deposition beyond the banks of entrenched channels during recent flooding. The depth of channel downcutting is restricted by a calcic soil horizon (>10,000 years old), which in places is highly cemented (Figure 4-10). Thus, the grade of the channels is controlled by developed calcic soil.

The longitudinal profile of the basin and alluvial fan is slightly concave (Figure 4-11), with a longitudinal slope of 0.016 at the toe and 0.019 at the apex. The relief ratios of the fan and drainage basin are 0.018 and 0.024, respectively, and are small for fans in Arizona (Hjalmarson and Kemna, 1991). The small relief ratio of the basin suggests that debris flows are unlikely.

The braided pattern of the distributary channels may have resulted when the old sediment was deposited, and the channel trenching possibly was initiated by postglacial climate change. Because the area is tectonically inactive (Rhoads, 1986), climate change is a plausible cause of the channel incision. There is evidence in other areas of channel entrenchment during a single storm, but the interlacing drainage pattern and rather uniform hydraulic geometry of the channels indicate that entrenchment and stabilization took place over a long period. Incision of Cave Creek, the base-level stream to the west, has resulted in minor tributary headcutting below the toe of the fan. The headcutting has little effect on the fan because the grade of the dominant distributary channels is controlled by resistant blocks of calcic soil above the Cave Creek channel. The interfluves are much wider near the toe, with little transverse relief except at the few incised distributary channels.

Accounts of Flooding

Meteorologic records and published accounts of major storms and floods indicate there have been at least three notable floods in the area during this century. The floods were on October 23, 1956, June 22, 1972, and October 6, 1993. Runoff and sediment information, furnished mostly by the Flood Control District of Maricopa County, is summarized for the second and third floods.

FIGURE 4-10 View looking upstream at distributary channel where the grade is controlled by a calcrete mass located above the backpack. The channel bed abruptly rises approximately 0.3 m at this spot. The outside dimension of the square frame is 1.5 feet (0.46 m). Courtesy of H. W. Hjalmarson.

Storm of June 22, 1972 Heavy rains in amounts of more than 100 mm (3.9 in) fell in nearby mountains within a 2-hour period. Although the storm center was to the south of Carefree fan, there was considerable runoff in the area. The peak discharge for Indian Bend wash, which drained 360 km^2 (138.8 m^2), of which Carefree fan is a part, was 595 m^3/s (21,015 feet^3/s) and is the highest peak since at least 1922. Unit peak discharges determined by the USGS for small mountainous basins to the south were from 5.77 m^3/s/km^2 (527.60 feet^3/s/mi^2) to an unusually large 37.2 km^3/s/km^2 (23.1 mi^3/s/mi^2). Unit peak discharge for the pediment above the Carefree fan undoubtedly was smaller because rainfall amounts were less and orographic effects were unlikely.

Photographs of floodflow and deposited sediment at road dip crossings suggest that there was a considerable amount of runoff and sediment movement in and through Carefree fan. Coarse sand bed material was deposited in road dip crossings along tributary streams at Scottsdale Road (Figure 4-9). Although dip crossings act as sediment traps as floodflow expands hydraulically (loses kinetic energy), the large deposits clearly show that large amounts of sediment moved onto the fan. There were no accounts of deposition in major areas of distributary channels on the

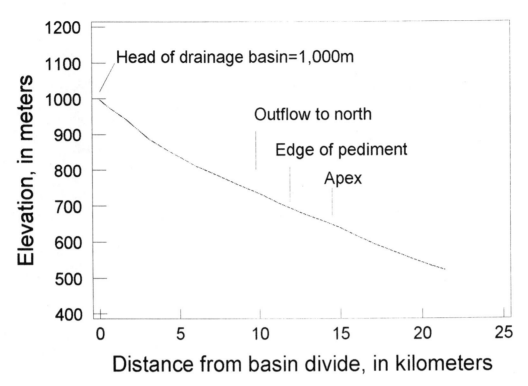

FIGURE 4-11 Profile of Carefree alluvial fan in Arizona.

piedmont plain, but there were large areas of sheetflow and undoubtedly deposition in lower parts of the piedmont plain, mostly below and to the south of Carefree fan. A few meters of bank erosion along distributary channel banks incised in younger deposits was observed. It is significant that there are no known accounts of flow path movement on the piedmont plain where the Carefree fan is located.

Storm of October 6, 1993 Tropical Storm Norma converged over central Arizona with a cold front associated with a strong Pacific low-pressure system and produced an average of 4.3 cm (1.7 in) of rainfall over Carefree fan and drainage basin. Flow was confined in nearly all distributary channels and flow divided and combined at all major channel forks and joins. Some sheet flooding and deposition in shallow overflow areas were observed in places near the fan toe. Measurements and estimates of peak discharge and runoff at 35 inflow, outflow, and internal channels show a peak discharge of 1.76 m^3/s (62.16 $feet^3$/s) in a single channel from the 10.9-km^2 (4.2 mi^2) drainage basin at Scottsdale Road and a total peak discharge of 13.0 m^3/s (459.2 $feet^3$/s) from 15 channels near the fan toe. Near the center of the fan, the total peak discharge at 11 tributary and distributary channels was 10.6 m^3/s (374.4 $feet^3$/s). These data show that a significant part of the runoff and peak discharge at the toe of Carefree fan was from the fan itself.

Storm runoff within the bounds of the Carefree fan and its drainage basin is directly related to total storm rainfall. The simple linear relation indicates approximately 41 mm (1.6 in) of rainfall were needed to produce runoff and the proportion of rainfall that entered the larger tributary and distributary channels increased with the amount of rainfall. For example, the relation indicates that for 4.3 cm (1.7 in) of rainfall only about 0.11 percent ran off but at a total storm rainfall of 4.6 cm (1.8 in) about 0.32 percent ran off.

Flow Path Movement

The tree-lined distributary channels indicate that the flow paths are fixed and in a condition of relative stability. Many of the large Palo Verde and mesquite trees along the channels are visible on aerial photographs taken on September 7, 1941, March 8, 1953, and March 30, 1991. A comparison of stream channels depicted on these good-quality large-scale aerial photographs revealed no change in the location of flow paths on Carefree fan. Rhoads (1986) also found no major changes in the form of channel networks in the general region for a 30-year historical period using several sets of aerial photography.

Characterizing Alluvial Fan Flooding Processes

Floodflow from the basin is conveyed through a long channel bounded by a narrow drainage basin to the apex. The drainage basin is not expected to produce large flood peaks because (1) the basin has a general "banana" shape with an average width of 770 m (2,530 feet), or only 5 percent of the basin length and (2) the first-order streams convey flow to only a few long second-order streams. Below the apex, floodflow enters the network of incised distributary channels located between stable ridges within the alluvial fan. The rainfall and runoff data for the storm of October 6, 1993, show that large amounts of discharge originated on the fan. The abnormally large fraction of the peak discharge that originated on the fan for that particular storm was partly because of an uneven rainfall distribution over the fan and basin. This suggests that the peak flood discharges on the lower part of the fan would be greater than predicted with normal application of the FEMA procedure (FEMA, 1990) as that procedure does not take into account runoff from the fan.

A simplifying flood characteristic is that avulsions, if any, are rare because there is little alluviation. Flood characteristics, however, are complex because (1) floodwater may be confined to defined channels or may inundate large portions of land between channels, (2) floodflow is generated from rainfall on the fan itself, (3) floodwater is lost to infiltration within channels, and (4) peak discharge is reduced by attenuation as floodflow entering the fan divides into channels and is temporarily stored on the fan surface. Also, distributary channels may be subject to scour and fill, while the interfluves separating the channels are composed of developed soils that resist lateral channel movement. Because flowpaths are confined by stable interfluves and there is little alluviation, there currently is no active flooding on Carefree fan.

Defining Areas of Active Alluvial Fan Flooding Hazard

There are no areas on the Carefree fan where flow paths are expected to change.

Where does flow depart from confined channels? Floodflow typically is confined within and adjacent to the trenched channels.

Where does sheet flow deposition occur? During the largest recent flood of October 6, 1993, there was minor overbank flooding at two low-lying areas of the lower fan. On this basis and because of the relatively small topographic relief, sheetflow with some sediment deposition might be expected during major floods in the 4.0-km^2 (2.5 mi^2) area defined as "sheetflow in low-lying areas" (Figure 4-9).

Where does debris flow deposition occur? There is no evidence of debris flows.

Where are there structures or obstructions that might cause or aggravate alluvial fan flooding? There are no structures or obstructions that might cause flow path movement. Some bank erosion in young soils may result at a few locations where structures are near the banks of distributary channels. Structures in geomorphically active areas (young soils along the distributary channels) should be avoided.

Where can the flood hazard be mitigated by means other than major structural flood control measures? Low-density development with restriction of structures to the stable ridges of old soils between the distributary channels and the elevation of structure floors form an effective means of mitigating the flood hazard.

Defining Areas of Nonalluvial Fan Flooding Hazard Along Stable Channels

In general, except during major floods, floodwater is confined to the entrenched channels. Flooding of approximately 6 percent of the 9.9-km^2 (3.8 m^2) alluvial fan area, mostly along the 30 km (18.6 mi) of defined channels, is estimated to have a probability of 1 percent in any year. Most of the discharge of the 100-year flood, estimated using methods of Thomas et al. (1994), will be conveyed in the channels with some shallow overbank flow commonly adjacent to the channels. Some overbank flow probably will combine with direct runoff of nearby tributary channels that have formed on the fan.

Determining the Types of Processes Occurring on the Active Parts of the Alluvial Fan

Recent unconsolidated deposits along the margins of the distributary channels are stratified. These young soils are composed of several alternating layers of silt, sand and small gravel and very thin layers of clay. There are no large unstratified deposits or large marginal levees of very coarse material associated with debris flows. Local small levee deposits of coarse sand and fine gravel on gently sloping channel banks were apparent at two sites following the October 6, 1993, flood. These small levees were less than 50 m (164 feet) long and are associated with the remobilization and movement of bed sediment of the distributary channels. Thus, water flood processes are dominant and channel deposits of coarse sand can remobilize as debris masses for short distances.

Using Soil Maps to Define the Nature of the Fan and the Extent of Flooding

NRCS soil survey reports with 7.5-minute orthophoto maps depicting types of soils are useful for rapid assessment of active and inactive areas on fans (Cain and Beatty, 1968). For example, the latest available map of soil units (Camp, 1986) shows that most of Carefree fan is a terrace with well developed soil profiles at fairly shallow depths: the lower part of the fan is Momoli soil and most of the middle and upper parts are Pinaleno soil (general soil map, Figure 4-9). Within these areas of predominantly old soils are units of predominantly young soils such as the Anthony-Arizo unit along the stream channel in the upper par of the fan. The young soils correspond to areas along stream channels where sediment may be subject to fluvial erosion and deposition or where horizons have not yet developed. Thus, significant low-cost information is gleaned from a cursory examination of the soil survey report, which shows that Carefree fan is not active.

To demonstrate the value of more detailed mapping of soil units, part of Carefree fan was recently mapped by the NRCS on aerial photographs at a scale of approximately 1:5,000 (Cathy E. McGuire, soil scientist, NRCS, written communication). The detailed mapping typically showed a close correlation between young soils and flood channels as defined using hydraulic methods (detailed soil map and flood map, Figure 4-9). Stable interfluves of older soil are also closely correlated with areas above defined flood levels. Thus, for Carefree fan the detailed soil surveys by the NRCS, which typically are not published, were useful but not essential for the assessment of flood characteristics.

RUDD CREEK, UTAH

Rudd Creek Canyon lies on the face of the western slopes of the Wasatch front in Davis County, Utah, north of Salt Lake City at the community of Farmington. The steep 1.8-km^2 (0.7 mi^2) basin rises about 1,150 m (3,773 feet) above the fan apex and is underlain by bedrock and covered with grasses, scrub oak, and mountain mahogany vegetation. The Wasatch fault, located near the mountain front, is near the apex of the fan.

A massive sedimentation event that consisted of several debris flows over the course of approximately a week originated in this steep canyon in the late spring of 1983 during rapid runoff from snowmelt. The debris flows, which included an especially large debris flow, caused damage to 35 structures on the fan; 4 homes near the topographic apex were completely destroyed, and 15 other homes were severely damaged. This event has been the subject of several published reports and has a substantial amount of documentation, including a video tape (Costa and Williams, 1984) of one of the debris flows. The video tape shows massive angular boulders being rapidly pushed or rafted down the steep canyon toward the alluvial fan.

The area is complex because the occurrence of large debris flows on Rudd Creek may be so infrequent as to be outside the realm of traditional hazard consideration. Also, debris flows result from the triggering or release of sediment that has accumulated to threshold levels over hundreds of years. Future debris flows probably will be smaller than the 1983 event until sediment again accumulates to threshold capacity and is released to the fan below. Clearly, assessment of debris accumulation in the drainage basin should be part of hazard mitigation as described by Keaton (1995).

Recognizing and Characterizing Alluvial Fans

Determining Whether or Not a Landform Is an Alluvial Fan

On the basis of the criteria established by the committee, Rudd Creek is an alluvial fan. Its composition, morphology, and location all fit the requirements, as detailed below.

Composition   The upper slopes are coarse debris fan deposits. The soil of the upper part of the fan is Kilburn cobbly sandy loam that occurs on short, slightly convex slopes of alluvial fans, mainly along the channel of intermittent streams (NRCS, 1968). The soils along the fan are characterized by the NRCS as occurring on alluvial fans in the uplands. The parent materials are alluvium and colluvium derived mainly from gneiss, quartzite, and granite.

Morphology   The site is shaped like a fully extended fan with a general convex shape and concentric contour lines. Locally irregular contour lines depict streets and other urbanization. The debris basin near the apex was constructed following the 1983 debris flow (Figure 4-12).

Location   The depositional zone begins at a topographic break associated with the Wasatch fault where the stream channel loses definition. The zone extends downslope to Farmington Creek, which forms the toe of the fan.

Defining the Lateral Boundaries and Topographic Apex of the Alluvial Fan

The fan is on a bajada and the lateral bounds generally are at swales where the fan coalesces with adjacent fans. Construction of a debris basin as well as urbanization on the fan has past altered the natural bounds. Inspection of the soil map indicates that alluvial deposits extend along the toe of the steep slopes equidistant from the mouth of the canyon. The topographic apex is readily apparent in Figure 4-12.

Defining the Nature of the Alluvial Fan Environment

Active Fan

Geologic assessment of sedimentation indicates that seven major debris flows have occurred during the Holocene (Keaton et al., 1988), with two of the flows occurring in the 140 years (historical period). About 63,000 m^3 (2.2 million feet^3) of sediment was deposited on the fan in 1983. Deposited debris in the upper part of the fan is remobilized and transferred downfan by water floods.

FIGURE 4-12 Topographic map on orthophoto base of Rudd Creek Canyon. The debris basin shown near the center of the map was built where homes were destroyed during the 1983 debris flow. Courtesy of Aero-Graphics, Inc.

Relict Fan

Although major debris flows have been infrequent during the past 10,000 years, the entire fan would be considered active.

Mountainous Drainage Basin

The steep, rugged drainage basin is heavily vegetated. This vegetation protects the watershed from erosion and landslides, which are a source of rock for debris flows. In the absence of wildfires and overgrazing, the accumulation of colluvium in the canyon is slow. Accumulation to threshold levels where heavy moisture can trigger movement of the colluvium and form major sediment events takes approximately 1,400 years. Because much of this colluvium was removed in 1983, and deposited on the alluvial fan, many years of accumulation may be needed before threshold levels are again reached. However, the frequency of major debris flows may be nonstationary because of changed land use practices.

The amount of sediment available for debris flows is related to the length of channels in the basin. When progressive sediment accumulation approaches the threshold that leads to sedimentation events on the fan, about 10 to 12 yards^3/foot (7.6 to 9.8 m^3/m) of channel length of colluvium is stored in the basins. This amount of unit storage is common for basins along the Wasatch front in Davis County (Sidney Smith, floodplain manager of Davis County, oral communication, 1995).

Sedimentation Event of 1983

The debris flows of 1983 were generated by mobilization of colluvium in the main channel of the basin as described above and by Lowe (1993) and Mathewson and Keaton (1990). This is the largest event recorded for the watershed since Davis County began observations in 1847. Similar snowmelt in 1984 produced very little debris in comparison.

Flow Path Changes

The 1983 debris flows obliterated the preflow paths of flow, but the paths of debris flow and subsequent floodflow were restricted by the urban development on the fan. For example, several homes blocked debris and shielded downslope areas.

Characterizing Alluvial Fan Flooding Processes

Defining Areas of Active Alluvial Fan Flooding Hazard

Where does flow depart from confined channels? The 1983 debris flows were not confined below the topographic apex as described above.

Where does sheetflood deposition occur? There appears to be little sheetflooding except in the midfan area where remobilized debris flow deposits are deposited by water flows.

Where does debris flow deposition occur? The entire fan might be subject to flooding if the newly constructed debris basin were overwhelmed by a large debris flow. The 1983 flows remained in the channel until they reached the apex. Sheetflow and deposition will occur downstream of the debris basin if it is filled and the outflow channel blocked. Studies by Keaton (1995) and Keaton et al. (1988) have determined that the debris basin should be able to hold all of the sediment from most events. Urbanization of the fan has altered drainage patterns and most flooding is likely to occur along the streets.

Alluvial fan flooding emanating from the canyon now will be confined by the debris basin and overflows will most likely be diverted by Farmington Canyon Road and other local streets.

Where are there structures or obstructions that might aggravate or cause alluvial fan flooding? Structures below the debris basin form obstructions, and streets become flow paths.

Where can the flood hazard be mitigated by means other than major structural flood control measures? Possible methods of mitigating debris flow hazards include avoidance, source area stabilization, transportation-zone (debris flow track between the source area and the depositional zone) modification, and defensive measures in the depositional zone (Lowe, 1993). Because of the many residential structures on the fan, a warning alarm might be used to avoid loss of life. Also, source area stabilization through maintenance of dense vegetation, including prevention of basin wildfires, can mitigate the flood hazard. Maintenance of the debris basin, a major structural control, is fundamental.

Determining the Type of Processes Occurring on the Active Parts of the Alluvial Fan

The approximately 30-foot-high (10-m-high) snout of Rudd Creek debris flow was low in clay and high in clast content suggesting a cohesionless flow with considerable intergranular collision (Shanmugan, 1996; Keaton et al., 1988). The video tape of a subsequent smaller flow shows large, rafted, angular boulders being rapidly pushed down the steep canyon. The actual peak water flow during this sedimentation event has been estimated to be 85 feet^3/s (2.4 m^3/s), a seemingly insignificant amount of flow. The fan is formed by about seven massive debris flows in the Holocene and many small debris flows. Deposited debris has been at several locations, including some above the apex. The coarse deposits have been remobilized and transported downfan, where some sediment is carried away at Farmington Creek at the fan toe.

HUMID REGION ALLUVIAL FANS

Alluvial fans and alluvial fan flooding are not limited to semiarid sites in the western United States; indeed, both are prevalent throughout the Appalachian Mountains of the eastern United States, from Tennessee to New Hampshire (Kochel, 1987). Although numerous landforms that would be described as alluvial fans exist in the Appalachian Mountains, many appear to be inactive or to have active parts that are much smaller than the overall landform. For example, an extensive apron of alluvial fan deposits mantles the western slopes of the Blue Ridge in Virginia, yet evidence to date seems to indicate that flooding has not occurred on any of the coalescing fans. In fact, the overlapping margins of the fans, where sediments are finest, are the sites of deeply incised channels. The stratigraphy and morphology of deposits on the now abandoned interior fan surfaces indicate that fluvial processes once crossed the fans. Some workers have attributed the large, relict landforms to past climates, and in particular to warmer and wetter conditions during Tertiary time (> 2 million years ago).

A common type of alluvial fan in the Appalachians has been characterized by episodic debris flow deposition throughout the Holocene epoch. These fans are very small (<1 km² (0.6 mi²)), thin (deposits 5 to 20 m (16.4 to 65 feet) thick), steep (>10° longitudinal profile), and elongate in comparison to arid region fans (Williams and Guy, 1973; Mills, 1983; Mills and Allison, 1994; Kochel, 1987). The characteristics of these fans can be attributed to the dominant fan-forming mechanism in the region: debris flows that develop from rainfall-generated debris avalanches emanating from small, low-order channels draining steep mountain slopes (Figure 4-13a and 4-13b) and to the fact that they have accumulated slowly, probably only during the Holocene. Storms that have triggered flooding and deposition on small debris fans have included Hurricane Camille (1969, Virginia); Hurricane Juan (1985, Virginia and West Virginia); and the June 1995 storms in Madison County, Virginia. Recent hurricanes have caused devastating results such as

- deep gullying along and upstream from fan apexes;
- debris slides and avalanches along steep channels that feed into small, steep fans; and
- substantial scour and deposition along larger channels into which steeper mountain streams deposit their water and sediment.

Population is growing in the East as elsewhere, and the alluvial plains surrounding low to moderately high mountains are prime spots for development. Developers consider the piedmonts to be safer than the valley floors, but during hurricanes their safety is dependent on the stability of slope deposits upstream from the piedmont. Most debris avalanches originate in colluvial hollows, or bowl-shaped concavities that collect sediment between debris-flushing events and concentrate subsurface water flow during storms (Figure 4-14). Although such sites could be identified and mapped by a professional geomorphologist, no attempt is usually made to locate these areas of potential catastrophe. As a result, homes have been built on debris fans downslope from hollows that are in the process of being charged with soil that could be released, perhaps during the next intense rainfall (see Figure 4-15 for an example of a home that was destroyed by a debris flow produced by evacuation of a colluvial hollow).

In the following pages, examples of several small debris fans in Nelson County, Virginia, are considered according to the procedure used by the committee for identifying areas of potential alluvial fan flooding[1] (Figure 4-16). Fans in this area have been studied and described by Williams and Guy (1973) and Kochel (1987). In 1969, during Hurricane Camille, about 50

---

[1] The figures and field data used in this discussion are actually a composite from several debris fans in the Lovingston area of Nelson County, rather than a single, small fan. The work of Kochel (1987) includes analysis of more than 1 dozen fans in west-central Virginia and concludes that all are very similar in their depositional histories and processes.

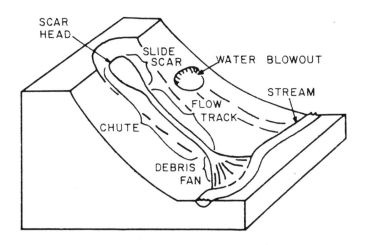

FIGURE 4-13 (Top) Block diagram of a debris avalanche that terminates on a debris fan. (Bottom) Debris avalanches that formed during Hurricane Camille, 1969, on hillsides terminating on two debris fans along the Virginia piedmont. East Branch of Hat Creek in lower left foreground removes sediment and water from the toes of the fans. Note that both debris fans are elongate, and active streams are found along the margins on the right in the photo. SOURCE: (Top) Reprinted with permission from Clark (1987). (Bottom) Williams and Guy (1973). Original photograph courtesy of Virginia Division of Mineral Resources (photographer T. M. Gathright II).

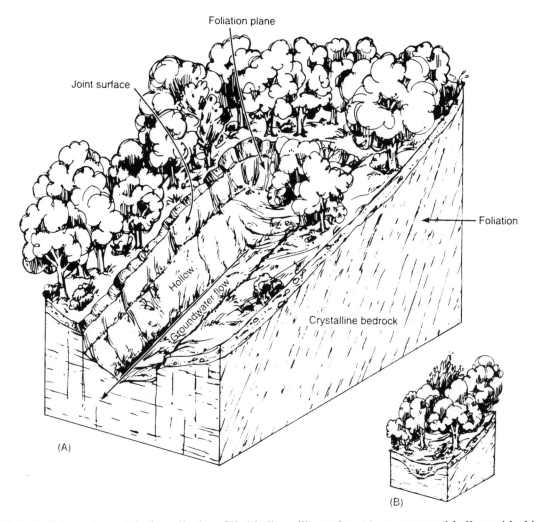

FIGURE 4-14 Schematic model of a colluvium-filled hollow, illustrating (a) an evacuated hollow with thin soils developed on hillsides and concentration of subsurface water flow in the trough of the hollow floor, and (b) the thick accumulation of colluvium and slope wash in the hollow after several hundred years. SOURCE: Reprinted with permission from Ritter et al. (1995).

percent of the surface area of many fans was flooded and altered by sedimentation from debris flows. As described below, these fans would be mapped and classified as potentially hazardous according to the procedures recommended in this document.

Recognizing and Characterizing Alluvial Fans

Determining Whether or Not a Landform Is an Alluvial Fan

Composition Examination of geologic maps alone does not provide enough information to identify the landform because the debris fans are not mapped separately from the bedrock mountain slopes; thus field work must be used to determine whether or not the landform is

FIGURE 4-15 (Top) The town of Lovingston, Virginia, is developed on a debris fan at the base of a debris avalanche that failed during Hurricane Camille on August 20, 1969. (Bottom) This clubhouse located on a debris fan was destroyed during a Blue Ridge, Tennessee, storm in 1973. At the time of the storm, the clubhouse was unoccupied. Upstream failure in the colluvial hollow resulted from failure of an access road fill built on the hollow. SOURCES: (Top) Williams and Guy (1973); photograph by Ed Rosenberry. (Bottom) Reprinted with permission from Clark (1987).

FIGURE 4-16   Debris avalanches and debris fans in Nelson County, Virginia, that formed during the 1969 hurricane. Courtesy of Virginia Division of Mineral Resources; photograph by T. M. Gathright II.

composed of alluvial sediments. Field studies of small debris fans in Nelson County indicate that the fans are composed solely of angular, very poorly sorted, mud matrix-supported gravels that range in size to as large as 5 m (16.4 feet) in maximum dimension (Kochel, 1987). Because these sediments are unconsolidated and have characteristics that indicate deposition from streams or debris flows, the composition of the landforms meets the criteria for the definition of an alluvial fan.

Morphology   To meet the criteria in the committee's definition of an alluvial fan, the landform of interest must have the shape of a fan, either partly or fully extended. Examination of a plane table map for one small fan in Nelson County (Figure 4-17) illustrates that the landform has the shape of a partly extended fan. Low-altitude photographs indicate that all of these piedmont landforms have an elongate fan shape.

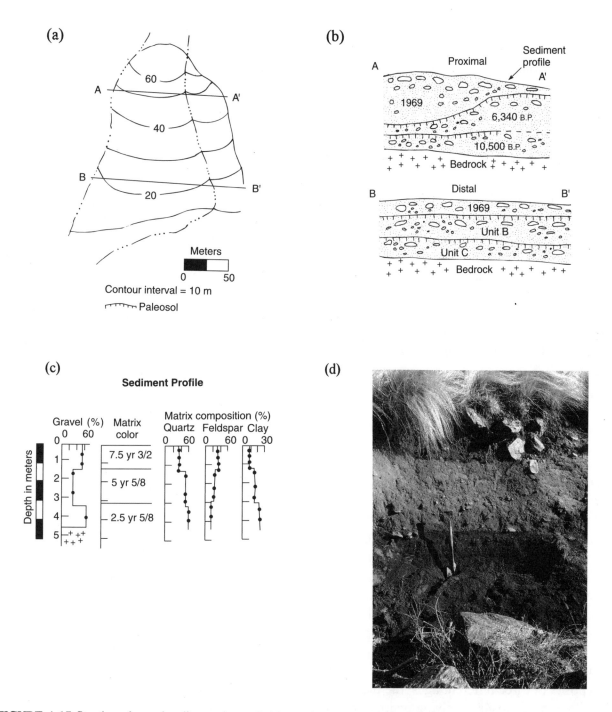

FIGURE 4-17 Stratigraphy and sedimentology of older and younger (1969) deposits on the Valentino debris fan. (a) Plane table map with locations of cross sections A-A' (proximal) and B-B' (distal). (b) Cross sections were measured and described in stream bank exposures and trenches dug on the fan surface. Radiocarbon ages indicate that older deposits are early Holocene and mid-Holocene in age. (c) Texture and matrix composition vary with depth, enabling the investigator to discriminate boundaries between debris flow events. These differences are illustrated in (c). (d) In this photo of a stratigraphic section, note the coarse, angular deposits from Hurricane Camille above the glove and two older deposits with paleosols indicated by the shovel. SOURCES: (a), (b), and (c) are reprinted with permission from Kochel (1987); (d) is reprinted with permission from Kochel and Johnson (1994).

Location To meet the criteria in the committee's definition of an alluvial fan, the landform of interest must be located at a topographic break where long-term channel migration and sediment accumulation become markedly less confined than upstream of the break. The Valentino fan, like other small debris fans in Nelson County, is located at a topographic break in slope along the eastern flank of the Blue Ridge Mountains.

Defining the Toe, Lateral Boundaries, and Topographic Apex of an Alluvial Fan

The gradients of the lower parts of the fans are gentler than those at the fan apexes, as can be seen from the greater spacing of contour lines in Figure 4-17a. The small debris fans in Nelson County (Figure 4-18) typically are encircled by slightly incised streams, many of which wrap around the toes of the fans (see also the shape of the ephemeral stream on the right in Figure 4-17a). These encircling streams, many of which are ephemeral, form the lateral boundaries and toes of the fans in Nelson County and can be identified on contour maps.

The topographic apex of the Valentino fan occurs approximately at the boundary between grass-covered slopes and tree-covered, steep hillsides. This is also the point where flow in the channel becomes unconfined and more uncertain and thus is coincident with the hydrologic apex.

Defining the Nature of the Alluvial Fan Environment and Identifying the Loci of Active Sedimentation

Defining Active

The committee recommends that the term *active* be used to refer to that time period during which sedimentation and flooding are possible in the current regime of climate and watershed conditions. In Nelson County, evidence is available to document when and how often episodic debris flow flooding and deposition have occurred. Furthermore, because it is clear that debris flows are associated with evacuation of hollows that must be full or close to full with colluvium before failure, it is possible for the investigator to have some understanding of the likelihood of activity on a given slope and its downstream depositional fan (Reneau et al., 1986).

In Nelson County, debris flow deposition and flooding have occurred about three times in the past 11,000 years, the Holocene (Kochel, 1987). Most fans have been constructed from deposits of three different ages, the oldest of which rests on late Pleistocene (~13,000 years B.P.) solifluction deposits overlying bedrock and has been radiocarbon dated at about 11,000 years at two sites. A younger deposit that has a mid-Holocene age (6,340 years B.P.; Figure 4-17b) is sandwiched between the basal unit and the historical deposit left by Hurricane Camille in 1969. From these data, geologists estimate that recurrence intervals for episodic debris flow deposition in the area are on the order of >3,000 to 4,000 years (Kochel, 1987), although from a stochastic hydrologist's perspective it is important to note that these events are not random in time.

Kochel (1987) provides evidence in support of the hypothesis that debris flow activity in the area was initiated by the Pleistocene-Holocene climatic transition, and in particular by the

FIGURE 4-18 A debris flow fan and recent drastically deposited rubble from the headwaters of the North Prong of Davis Creek that formed during Hurricane Camille in 1969. SOURCE: Williams and Guy (1973). Original photograph courtesy of the Virginia Division of Mineral Resources (photographer T. M. Gathright II).

onset of incursion of tropical air masses and moisture into the central Appalachians. Incursion of tropical moisture was concurrent with retreat of the polar front as Pleistocene glacial conditions waned and ice masses withdrew northward. These moist air masses can become locked in the steep, rugged terrain of the central Appalachians, unleashing intense rainfall events over short time periods. Many workers have documented that some colluvial hollows throughout the region are primed and prone to failure during such storms (Hack and Goodlett, 1960; Clark, 1987; Kochel, 1987; Jacobson, 1993).

For all the reasons summarized here, it seems evident that the past 11,000 years has been a time of active albeit sporadic debris fan growth and development in west-central Virginia. If no historical flooding had occurred, and the youngest deposits throughout the region were mid-Holocene in age, an investigator might be tempted to consider only the past few thousand years as critical to assessing the potential for flooding. However, historical flooding and fan sedimentation have occurred, and all deposits indicate long recurrence intervals that probably reflect the amount of time necessary to replenish the sediment supply in colluvial hollows. As a consequence, in this case a time period of 11,000 years is chosen as the best representation of whether or not a fan is active. (One must keep in mind, however, that numerous alluvial fans on the western flanks of the Blue Ridge have no historical or Holocene deposition and thus would be mapped as completely inactive according to this choice of time unit.)

An additional reason for this choice of time unit is the possibility that human activities now increase the potential of evacuation of colluvial hollows. As both tourism and urbanization are increasing in the region, it is probable that the potential for activity on a debris fan has increased.

Identifying Areas of Flooding and Deposition for the Time Period Chosen to Represent the Active Part of an Alluvial Fan

The alluvial fans in Nelson County are very small, and mapping by Kochel (1987) has indicated that single deposits can be traced across the entire fanhead area (Figure 4-17b). In addition, photographs of flooding during Hurricane Camille indicate that areas comprising up to 50 percent of the total fan area were flooded.

Finally, some photographs indicate that human-engineered structures and developments affected the paths of flow on mid- and lower fan areas. Therefore, it is prudent to map the entire fan as active, unless it can be demonstrated that part of a fan is of such high relief and resistance relative to the channelways that it is unlikely to be affected. If a fan did not have flooding and sedimentation during Hurricane Camille, it may be more likely to be flooded during the next large storm because upslope colluvial hollows have not been evacuated in several thousand years.

Defining and Characterizing Areas of Alluvial Fan Flooding on Active Parts of Alluvial Fans

Defining Areas of Alluvial Fan Flooding Hazard

Identifying areas where flow departs from confined channels (i.e., where flow paths are uncertain) For the same reasons as stated earlier, all parts of the small fans in Nelson County appear to be susceptible to alluvial fan flooding. As can be seen in Figures 4-13b, 4-15, 4-16 and 4-18, flow paths are highly unpredictable and prone to expansion and shifting. Because of the small size and low relief of the fan surfaces, channel migration is possible on any part of the fan.

Identifying areas where sheetflood deposition occurs Sheetflood deposition can be seen commonly at the toes of the fans in Nelson County (see Figures 4-13b and 4-16), especially where the fans grade into floodplains of larger streams, and thus have very low slopes.

Identifying areas where debris flow deposition occurs In Nelson County, all deposition except the sheetflood deposits on the lower parts of the fans is the result of debris flows, as demonstrated by analyses of stratigraphy in the fans (Williams and Guy, 1973; Kochel, 1987).

Identifying areas where structures or obstructions might aggravate or cause alluvial fan flooding Evidence of structural controls that aggravated or caused alluvial fan flooding in Nelson County can be seen in Figure 4-19, where an access road with fill built across a colluvial hollow might have triggered collapse of sediment in the hollow because of water collection and diversion. Outside of Nelson County, along the Potomac River where other examples of humid region alluvial fans occur, a good example of the migration of channel flow along a highway can be seen (Figure 4-20).

Defining Areas of Nonalluvial Fan Flooding Hazard Along Stable Channels

None of the channels on the alluvial fans in Nelson County appear to be stable; thus all flooding is deemed to be alluvial fan flooding.

SUMMARY

Judging whether alluvial fans are subject to flooding is not necessarily simple. The intent of these examples is to illustrate and inform, not to second guess past decisions. Four out of the seven examples fit the revised definition of *alluvial fan flooding*, (i.e., Henderson, Thousand Palms, Rudd Creek, Nelson County). However, although Lytle Creek, Tortolita, and Carefree are alluvial fans, they do not meet the active fan criteria and are therefore not subject to alluvial fan flooding even though they exhibit some characteristics that distinguish them from riverine flooding. Lytle Creek illustrates a site where a large, wide trench cuts the fan from the topographic apex to Cajon Creek along the distal east side of the fan, thereby isolating adjacent parts of the fan from active sedimentation and alluvial fan flooding. Carefree illustrates a site where the network of incised distributary channels transport sediment through the fan and the fan presently is inactive. The dissected, coalesced fan remnants along the west side of the Tortolita Mountains function to transport sediment, are not aggrading, and do not have the shape of an alluvial fan. The seven sites illustrate the complex nature of flooding on alluvial fans and piedmonts and the advantage of using a systematic approach to flood hazard assessment such as suggested by this committee.

FIGURE 4-19 Human activities can increase the potential for release of materials, as shown by this mass movement initiated along an access road in an area damaged during the Blue Ridge, Tennessee, storm of 1973. SOURCE: Reprinted with permission from Clark (1987).

Applying the committee's definition to field examples reveals that there is a choice to be made between having a very inclusive definition that consists of cases such as alluvial fans, deltas, and braided alluvial washes; or a somewhat exclusive definition that leaves out all nonalluvial fan cases along with those that display certain characteristics but not others. The first alternative would include sites like the Tijuana River (Box 4-1 and Figure 4-21) and involves changing the name *alluvial fan flooding* in the current regulations to *uncertain flowpath flooding* and adopting the committee's definition thereof. If the definition is to include cases that fit only one or two criteria, the distinction between alluvial fan and nonalluvial fan must be dropped because it would be inconsistent to include (1) less severe situations where fans are inactive and not subject to the committee's definition of alluvial fan flooding (i.e., and thus not include alluvial fans such as Lytle Creek and Carefree) and (2) nonalluvial fans with stable but distributary paths of flow like much of the western slope of the Tortolita Mountains. This will result in a more inclusive definition. The second alternative is to keep the term *alluvial fan flooding* in the regulations, clarify that it applies only to alluvial fans, and adopt the committee's definition thereof. This will result in a more exclusive definition.

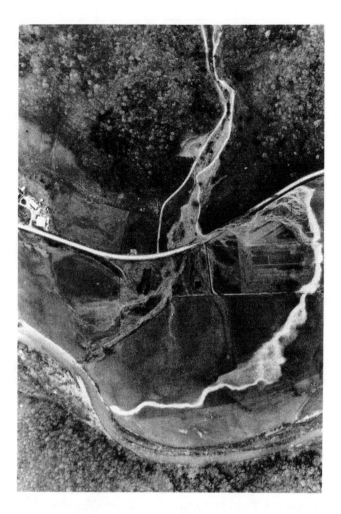

FIGURE 4-20 Debris fan at mouth of Nelson Run, tributary to North Fork South Branch Potomac River, Virginia. Flow along North Fork is from right to left; field of view is 750 m (2,461 feet) wide. Note diversion of tributary flow along road (to the right along base of mountain slope), which then moved back across the fan to reach the North Fork. SOURCE: Jacobson (1993).

This choice is to be made by FEMA based on policy, the tolerance for uncertainty in NFIP mapping procedures, and the resources available to restudy those areas that might fit a more inclusive definition. Regardless, the selection of either alternative will result in a better, more precise definition that can be directly applied to the physical processes associated with a given flooding source.

BOX 4-1
WHEN IT IS NOT A FAN BUT IT ACTS LIKE ONE:
TIJUANA RIVER, CALIFORNIA

The Tijuana River flows to the Pacific Ocean in San Diego County, California. Most of the 1,700-square-mile drainage area is in Mexico. Figure 4-21 was taken shortly after a large flood in 1993 and illustrates the effects of sediment deposition during a flood, which caused the river to split into more than one channel. This process resulted in an uneven distribution of flow across the floodplain and significant deviations from predicted base flood elevations.

The Tijuana is a coastal stream and clearly not an alluvial fan. However, this flooding source exhibits the characteristics of the existing NFIP definition of alluvial fan flooding: uncertain flow paths, high velocity flows that create new flow paths through erosion, shallow flow depths, and high sediment transport rates, which cause significant damage by depositing sand in and around structures. Applying the committee's proposed definition and field indicators to this case yields the following findings:

- Is it an alluvial fan? No. The Tijuana River is a coastal stream grading into a coastal delta.

- Nature of the fan environment: There is a perennial low-flow channel that presently appears to be the main channel. Riparian vegetation is present both in the channel and on other parts of the floodplain.

- Sediment transport: Coarse sediment yielded by watershed erosion processes has deposited long before it reaches this part of the river, which has a slope of approximately 0.2 percent.

- Topographic confinement: The floodplain is topographically bounded on the north and south. Due to natural flooding processes, however, there is considerable uncertainty associated with how flow paths are laterally distributed during a flood.

- Characterizing flooding processes: Review of historical floods since the mid-19th century indicates that the low-flow channel has shifted its location numerous times. Early maps also indicate that there used to be several low-flow channels along the floodplain. Flows do not spread evenly across the floodplain but rather form a number of concentrated conveyance regions. Application of the traditional flood paradigm (using HEC-2) has been somewhat successful even though it fails to account for bank erosion, avulsion, scour, and other aspects of the actual flooding processes.

- Structural mitigation required: Floodway setbacks, elevation on fill, or similar measures may not necessarily be adequate to mitigate the flood hazard.

Although this flooding source is topographically confined, natural sediment transport processes introduce considerable uncertainty into the prediction of stage-discharge relationships for the Tijuana River. The presence of multiple flow paths and the uncertainty they introduce does not, by itself, mean that this area is subject to alluvial fan flooding because the area is not an alluvial fan. The Tijuana River floodplain does, however, fall into the category of uncertain flow path flooding. It would seem reasonable that more stringent rules might apply to the mitigation of flood hazards for this case.

FIGURE 4-21 Tijuana River north of the Mexico border (1993). Courtesy of Aerial Fotobank, Inc.

REFERENCES

Bryant, B. 1988. Geology of the Farmington Canyon complex, Wasatch Mountains, Utah. U.S. Geological Survey Professional Paper 1476. Washington, D.C.: Department of the Interior.

Cain, J. M., and M. T. Beatty. 1968. The use of soil maps in the delineation of flood plains. Water Resources Research 4(1):173-182.

Camp, P. D. 1986. Soil Survey of Aguila-Carefree Area, Parts of Maricopa and Pinal Counties, Arizona. U.S. Soil Conservation Service Report. Washington, D.C: Department of the Interior.

Chin, E. H., B. A. Aldridge, and R. J. Longfield. 1991. Floods of February 1980 in Southern California and Central Arizona. U. S. Geological Survey Professional Paper 1494. Washington, D.C.: Department of the Interior.

Clark, G. M. 1987. Debris slide and debris flow historical events in the Appalachians south of the glacial border. Pp. 139-155 in Reviews in Engineering Geology, Volume VII. Boulder, Colo.: The Geological Society of America.

Costa, J., and G. Williams. 1984. Debris-flow dynamics. U. S. Geological Survey Open-File Report 84-606, 22-1/2 minutes (videotape). Reston, Va.: U.S. Geological Survey.

DMA Consulting Engineers. 1985. Alluvial Fan Flooding Methodology—An Analysis. Federal Emergency Management Agency Contract EMW-84-C-1488. Washington, D.C.: FEMA.

Eckis, R. 1928. Alluvial fans of the Cucamonga district, Southern California. Journal of Geology 36:224-247.

Federal Emergency Management Agency (FEMA). 1990. Fan, An Alluvial Fan Flooding Computer Program. Washington, D.C.: FEMA.

Fields, J. J. 1994. Surficial Process, Channel Change, and Geological Methods of Flood-Hazard Assessment on Fluvially Dominated Alluvial Fans in Arizona. Ph.D. dissertation, University of Arizona.

Fuller, J. E. 1990. Misapplication of the FEMA alluvial fan model: a case history. Pp. 367-377 in Hydraulics and Hydrology of Arid Lands, R. H. French, ed. New York: American Society of Civil Engineers.

Hack, J. T., and J. C. Goodlett. 1960. Geomorphology and Forest Ecology of a Mountain Region in the Central Appalachians: U.S. Geological Survey Professional Paper 347. Reston, Va.: U.S. Geological Survey.

Hereford, R., K. S. Thompson, K. J. Burke, and H. C. Fairley. 1995. Late Holocene Debris Fans and Alluvial Chronology of the Colorado River, Eastern Grand Canyon, Arizona: U.S. Geological Survey Open-File Report 95-97. Reston, Va.: U.S. Geological Survey.

Hjalmarson, H. W., and S. P Kemna. 1991. Flood Hazards of Distributary-Flow Areas in Southwestern Arizona. U.S. Geological Survey Water Resources Investigations Report 91-4171. Reston, Va.: U.S. Geological Survey.

Hooke, R. L. 1967. Processes on arid-region alluvial fans. J. Geol. 75:438-460.

Jacobson, R. B., ed. 1993. Geomorphic studies of the storm and flood of November 3-5, in the Upper Potomac and Cheat River basins in West Virginia and Virginia: U.S. Geological Survey Bulletin 1981. Reston, Va.: U.S. Geological Survey.

Keaton, J. R. 1995. Dilemmas in regulating debris-flow hazards in Davis county, Utah. Pp. 185-192 in Environmental and Engineering Geology of the Wasatch Front Region, W. R. Lund, ed. Publication 24. Salt Lake City: Utah Geological Association.

Keaton, J. R., L. R. Anderson, and C. C. Mathewson. 1988. Assessing Debris Flow Hazards on Alluvial Fans in Davis County, Utah. Final report for U.S. Geological Survey Landslide Hazard Reduction Program, Agreement No. 14-08-0001-A0507. Reston, Va.: U.S. Geological Survey.

Kochel, R. C., and Johnson, R. A. 1984. Geomorphology and sedimentology of humid-temperate alluvial fans, central Virginia. Pp. 109-122 in Gravels and Conglomerates, E. Koster and R. Steel. Canadian Society of Petroleum Geologists Memoir 10.

Kochel, R. C. 1987. Holocene debris flows in central Virginia: Pp. 139-155 in Reviews in Engineering Geology, Volume VII. Boulder, Colo.: The Geological Society of America.

Lowe, M. 1993. Debris-Flow Hazards: A Guide for Land-Use Planning, Davis County, Utah. U.S. Geological Survey Professional Paper 1519. Reston, Va., U.S. Geological Survey.

MacArthur, R. C. and D. L. Hamilton. 1986. Corps of Engineers efforts in modeling mudflows. Pp. 549-554 in Proceedings of a Western State High Risk Flood Areas Symposium, Improving the Effectiveness of Floodplain Management in Arid and Semi-Arid Regions. Madison, Wisc.: Association of State Floodplain Managers.

Machette, M. N. 1985. Calcic soils of the southwestern United States. Geological Society of America Special Paper 203. Boulder, Colo.: The Geological Society of America.

Mathewson, C. C., and J. R. Keaton. 1990. Multiple phenomena of debris-flow processes: A challenge for hazard assessments. Pp. 549-554 in Proceedings of the International Symposium, Hydraulics/Hydrology of Arid Lands. New York: Hydraulics Division of American Society of Civil Engineers.

McGlashan, H. D., and F. C. Ebert. 1918. Southern California floods of January 1916. U.S. Geological Survey Water-Supply Paper 426. Reston, Va.: U.S. Geological Survey.

Mills, H. H. 1983 Piedmont evolution at Roan Mountain, North Carolina. Geografisika Annaler 65A:111-126.

Mills, H. H., and J. B. Allison. 1994. Controls on the variations of fan-surface age in the Blue Ridge Mountains of Haywood County, North Carolina. Physical Geography 156:465-480.

Natural Resources Conservation Service (NRCS). 1968. Soil Survey of Davis-Weber Area, Utah. U.S. Department of Agriculture. Washington, D.C.: U.S. Department of Agriculture.

Natural Resources Conservation Service (NRCS). 1973. Soil Survey of San Diego County, California. Washington, D.C.: U.S. Department of Agriculture.

Natural Resources Conservation Service (NRCS). 1980a. Soil Survey of Riverside County, California, Coachella Valley Area. Washington, D.C.: U. S. Department of Agriculture.

Natural Resources Conservation Service (NRCS). 1980b. Soil Survey of San Bernardino County—Southwestern Part, California. Washington, D.C.: U.S. Department of Agriculture.

Patton, P. C., and V. R. Baker. 1976. Morphology and floods in small drainage basins subject to diverse hydrogeomorphic controls. Water Resources Research 12(5):941-952.

Pearthree, P. A., K. A. Demsey, J. Onten, K. R. Vincent, and P. K. House. 1992. Geomorphic Assessment of Fluvial Behavior and Flood-Prone Areas on the Tortolita Piedmont, Pima County, Arizona. Arizona Geological Survey Open-file Report 91-11. Tucson, Ariz.: Arizona Geological Survey.

Reneau, S. L., W. E. Dietrich, D. J. Donahue, A. J. T. Juli, and M. Rubin. 1986. Late Quaternary History of Colluvial Deposition and Erosion in Hollows, Central California Coast ranges. Geological Society of America Bulletin 102(7):969-982.

Rhoads, B. L. 1986. Flood hazard assessment for land-use planning near desert mountains. Environmental Management 10(1):97-106.

Ritter, D. F., R. C. Kochel, and J. Miller. 1995. Process Geomorphology, 3rd ed. Dubuque, Iowa, Times Mirror Higher Education Group.

Ritter, J. B., J. R. Miller, Y. Enzel, S. D. Howes, G. Nadon, M. D. Grubb, K. A. Hoover, T. Olsen, S. L. Reneau, D. Sacks, C. L. Summa, I. Taylor, K. C. N. Touysinhthiphonexay, E. G. Yodis, N. P. Schneider, D. F. Ritter, and S. G. Wells. 1993. Quaternary evolution of Cedar Creek alluvial fan, Montana. Geomorphology 8:287-304.

San Diego County. 1977. Storm Report for Tropical Storm Doreen, August 15-17, 1977. San Diego: County of San Diego Department of Sanitation and Flood Control.

Shanmugan, G. 1996. High-density turbidity currents: Are they sandy debris flows? Journal of Sedimentary Research 66(1):2-10.

Sharp, R. V. 1972. The Borrego Mountain Earthquake of April 9, 1968. U.S. Geological Survey Professional Paper 787. Reston, Va.: U.S. Geological Survey.

Thomas, B. E., H. W. Hjalmarson, and S. D. Waltemeyer. 1994. Methods for Estimating Magnitude and Frequency of Floods in the Southwestern United States. U.S. Geological Survey Open-File Report 93-419. Reston, Va.: U.S. Geological Survey.

Troxell, H. P. 1942. Floods of March 1938 in Southern California. U.S. Geological Survey Water-Supply Paper 844. Reston, Va.: U.S. Geological Survey.

Waananen, A. O. 1969. Floods of January and February 1969 in Central and Southern California. U.S. Geological Survey Open-File Report. Reston, Va.: U.S. Geological Survey.

Wells, S. G., and A. M. Harvey. 1987. Sedimentologic and geomorphic variations in storm-generated alluvial fans, Howgill Fells, northwest England. Geological Society of America Bulletin 98:182-198.

Williams, G. P., and H. P. Guy. 1973. Erosional and depositional aspects of Hurricane Camille in Virginia, 1969. U.S. Geological Survey Professional Paper 804. Reston, Va.: U.S. Geological Survey.

Williams, S. R., and M. Lowe. 1990. Process-based debris-flow prediction method. Pp. 66-71 in Hydraulics/Hydrology of Arid Lands, R. H. French, ed. New York: American Society of Civil Engineers.

5

Conclusions and Recommendations

The Committee on Alluvial Fan Flooding was charged to study how to improve the way we address alluvial fan flood hazards in the context of the National Flood Insurance Program (NFIP). Specifically, the committee was asked to develop an updated definition of alluvial fan flooding, to specify criteria that can be assessed to determine if an area is subject to alluvial fan flooding, and to provide examples that use the definition and criteria. The committee also endeavored to shed light on the conflict that has been associated with implementing the NFIP in areas with alluvial fans.

FEMA, as a federal agency, has great influence over the way communities manage and mitigate flood hazards. This influence comes both from its congressional mandate and from its discretion to withhold certain benefits from communities who violate NFIP regulations. It is a complex matter to regulate alluvial fan flood hazards, particularly on a national scale. In addition to the diversity of the flood hazards themselves, FEMA must deal with a wide range of communities, some of which do not have the resources for a technically sophisticated floodplain management program. Part of FEMA's leadership responsibility, however, is to set a consistent example.

The conclusions and recommendations in this chapter apply to alluvial fans and flooding that takes place on them. As discussed in Chapter 1, the problems of erosion and deposition processes, flow path uncertainty, and flood hazard severity are not phenomena limited to alluvial fans. Hence, to apply the term *alluvial fan flooding* to all such phenomena—as is now done—is confusing for those cases that do not actually occur on alluvial fans. This limitation needs to be noted explicitly, especially in regard to the committee's proposed definition of alluvial fan flooding as presented in Recommendation 1.

Another important point is that all parts of all alluvial fans are not inherently hazardous locations. Hazardous flood processes can, in some cases, occur over entire alluvial fans or, as is more common, only on the active portions of a fan. Relict alluvial fans are entirely inactive. Thus, to identify the location of alluvial fan flooding, flooding processes need to be understood in regard to their temporal and spatial relationship to the landscape. Although it is possible to create a general definition of alluvial fan flooding to meet the needs of regulators and planners, in reality each site must be evaluated individually to determine its specific character.

CONCLUSIONS

1. Site investigation is essential to distinguish alluvial fans from other landforms and to identify which parts of a particular alluvial fan are subject to hazard. Alluvial fans are only one of a variety of landforms that may make up a piedmont environment. Others include pediments, rock fans, and alluvial plains. In addition, alluvial fans themselves can range from untrenched active fans to fully trenched fans with large inactive portions outside of the channel. Between these ends of the fan continuum, there is a range of morphologies. Because of the variety of piedmont landforms and the variability of alluvial fan morphology, to determine flooding potential it is necessary for an experienced person to visit each fan to identify the landform type and the active and inactive components of the landscape.

2. Regulatory flexibility is necessary to effectively manage for flood hazards given the variability in flood processes on alluvial fans and alluvial fan morphology. Flood processes and characteristics on alluvial fans occur across a continuum that defies simple designations. Two of the major types of floods in this continuum are debris flows and water flows. There is also a continuum between active and relict fan surfaces, as influenced by long-term changes in tectonism and climate. Against this context of real-world processes and change, it becomes clear that any definition of a particular class of flooding is somewhat arbitrary and achieves precise meaning only through the establishment of regulations. Of course, these reflect natural processes imperfectly, at best. Although a definition may be necessary to achieve regulatory goals, any definition should be subject to review and revision in light of experience gained as the definition is applied in different locales.

3. The existing regulatory framework, which divides all flooding sources into riverine or alluvial fan flooding, leads to inconsistency when imposed on specific sites. The FEMA definition of alluvial fan flooding, as contained in section 59.1 of the National Flood Insurance Program (NFIP) regulations, cannot be separated from the broader flood-hazard context, including rule making, regulatory imperatives, and the need for administrative consistency and application nationwide. The image of alluvial fan flooding relevant to the NFIP definition is that of a type or class of flooding that is more uncertain, more dangerous, and therefore more hazardous than the ordinary flooding mapped on the Flood Insurance Rate Maps (FIRM). The FEMA definition of alluvial fan flooding applies to this general concept, rather than to the complex reality of the flood processes that occur on various individual alluvial fans. The mismatch between the general image and various individual cases leads to confusion and misunderstanding by those regulated according to local FIRMs that are predicated on the definition. The definition recommended in this report, and particularly its requirement of site-specific study, is intended to resolve the present confusion in distinguishing between riverine flooding, uncertain flow path flooding, and alluvial fan flooding.

4. Imposing the alluvial fan flooding paradigm instead of the riverine flooding paradigm creates its own set of difficulties for sound regulation of the hazard. In the present regulatory environment, given the intent of Congress for the NFIP, any definition of alluvial fan flooding, including the one recommended herein, has major consequences for flood hazard management. At a minimum, the definition will result in the invocation of special regulatory oversight by FEMA (with implications for the FIRM mapping procedures, local flood hazard management plans, and the expenses of mitigation measures) and in potential difficulties with reducing flood uncertainty that are not encountered with the riverine designation. Given its necessary context and its consequences, the required definition of alluvial fan flooding must allow for consideration of (a) the nature of the physical processes that actually occur on alluvial fans and the changes in processes and landforms that occur on fans through time, and (b) the concerns of current regulatory practice.

5. The act of defining flooding processes and characteristics is independent of the decision as to which methods are applicable for delineating the boundaries of the hazard. The method of hazard delineation developed by Dawdy (1979) and adopted by FEMA was specifically intended to be applied in areas that were obviously fan-shaped. Although Dawdy's proposed solution is a very special case applicable only to a subset of alluvial fans, the underlying approach of using the conditional probability equation is sound and quite general. This approach is applicable to many flooding situations that contain elements of uncertainty. For urbanized fans or regions that are not fan-shaped, an alternative solution to the conditional probability equation is necessary in order to map the flood hazard. The applicability of the definition for alluvial fan flooding is based on the characteristics of the flood hazard and not the method chosen to delineate the boundaries of flooding.

6. The role of uncertainty in mapping alluvial fan flood hazards is different from that for floodplain management and mitigation. Alluvial fan flooding has implications for floodplain management. When a flood hazard is delineated on an alluvial fan using the default assumptions in the FEMA guidelines, the resulting map is an expression of uncertainty rather than an indication of how a flood might occur. It is therefore of limited use for the mitigation and management of flood hazards. In this case, if the FIRM is interpreted literally, then it can be argued that any effort at mitigation short of complete channelization increases the flood risk on another part of the fan and may therefore be in violation of NFIP regulations. Giving floodplain managers the peculiar responsibility of preserving uncertainty would be an inappropriate use of the FIRM because mitigation of flood hazards should strive to reduce uncertainty. This will become more visible if FEMA decides to extend the alluvial fan flooding concept to nonalluvial fan areas.

RECOMMENDATIONS

1. The existing NFIP definition of alluvial fan flooding should be revised. The Committee on Alluvial Fan Flooding proposes the following definition and supporting explanation, which are to be applied only with due consideration to all other conclusions and recommendations that are made in this report:

Definition Alluvial fan flooding is a type of flood hazard that occurs only on alluvial fans. It is characterized by flow path uncertainty so great that this uncertainty cannot be set aside in realistic assessments of flood risk or in the reliable mitigation of the hazard. An alluvial fan flooding hazard is indicated by three related criteria: (1) flow path uncertainty below the hydrographic apex, (2) abrupt deposition and ensuing erosion of sediment as a stream or debris flow loses its competence to carry material eroded from a steeper, upstream source area, and (3) an environment where the combination of sediment availability, slope, and topography creates an ultrahazardous condition for which elevation on fill will not reliably mitigate the risk.

Supporting Explanation Alluvial fan flooding begins to occur at the hydrographic apex, which is the highest point where flow is last confined, and then spreads out as sheetflood, debris slurries, or in multiple channels along paths that are uncertain. The hydrographic apex may be at or downstream of the topographic apex. Such flooding is characterized by sufficient energy to carry coarse sediment at shallow flow depths. The abrupt deposition of this sediment or debris strongly influences hydraulic conditions during the event and may allow higher flows to initiate new, distinct flow paths of uncertain direction. Also, erosion strongly influences hydraulic conditions when flood flows enlarge the area subject to flooding by undermining channel banks or eroding new paths across the unconsolidated sediments of the alluvial fan. Flow path uncertainty is aggravated by the absence of topographic confinement or by the occurrence of erosion and deposition processes. Together, these characteristics create a flood hazard that can be reliably mitigated only by the use of major structural flood control measures or by complete avoidance of the affected area.

The potential for erosion and deposition, the related uncertainty in flow path behavior, and the imprudence of increasing elevation by filling an area as a mitigation measure are joint and separate characteristics shared among many flood hazards on depositional environments other than alluvial fans, although not usually with the same intensity. It stands to reason that some of the same rules should be applied to this more inclusive type of flood hazard, termed *uncertain flow path flooding*, as applied to alluvial fan flooding. Flood hazards that meet only one or two of the criteria in the definition make up this third category.

2.   In revising its definition, FEMA will need to choose between an inclusive definition and an exclusive definition. Applying the committee's proposed definition to examples reveals that there are two alternative ways in which FEMA can adopt it. Based on policy, the tolerance for uncertainty in NFIP mapping procedures, and the resources available to restudy currently mapped areas, FEMA will need to choose between having a very inclusive definition (i.e., a definition that includes alluvial fans, coastal streams, and braided alluvial rivers) or a somewhat exclusive definition that does not apply to any nonalluvial fan cases or cases that display certain characteristics but not others. The inclusive approach merely involves changing the term *alluvial fan flooding* in the current regulations to *uncertain flowpath flooding* and adopting the committee's definition with editing to the first sentence to indicate what is covered. If the definition is to include cases that fit only one or two criteria, the distinction between alluvial fan and nonalluvial fan must be dropped because it would be inconsistent to exclude the more severe situations that meet all three criteria merely because they do not occur on an alluvial fan. This will result in a more inclusive definition. The second alternative is to keep the term *alluvial fan flooding* in the regulations, clarify that it applies only to alluvial fans, and adopt the committee's definition. Separate regulations (or policy statements) will be necessary for other cases.

3. Site-specific process evaluation is the key to determining which alluvial fans and parts of alluvial fans are subject to flood hazards. Everyone involved in assessing the hazards associated with floods on alluvial fans should recognize that both the landform and the processes are highly variable in time and space. Their properties and characteristics cannot be generalized from published descriptions and applied to new field sites. Flooding on alluvial fans can only be evaluated on a site-specific basis. On-site evaluation of the flood hazard requires field investigations by specialists experienced in the scientific study of alluvial fan processes and the geomorphologic indicators of their present and past operation. Such on-site investigation is critical to provide a scientifically sound basis for hazard delineation and regulation. Hazard delineation done in the absence of such study should be held as provisional until proper field investigation by qualified specialists is accomplished. As outlined in Chapter 3 of this report, the field investigations required to support the revised definition will include a program with the following three elements:

(a) identifying the alluvial fans, including their boundaries, apex relationships, and setting;

(b) determining the alluvial fan environment, including general processes, incision state, and active or relict status; and

(c) characterizing the alluvial fan flood processes, including their sources, extent, erosion, and sedimentation.

4. Uncertainty in flood hazards should be evaluated directly rather than assuming it is either nonexistent or random. Appendix 5 of FEMA (1995) document No. 37, *Guidelines and Specification for Study Contractors*, should recognize that the present default assumption—which assumes complete uncertainty when the geologic feature is an alluvial fan—is seldom the best starting point for a realistic and usable assessment of the flood hazard. There are, however, techniques and information that can be used to deal directly with the uncertainty, particularly given knowledge to be derived from a site-specific geomorphologic study. Some preliminary suggestions in this regard are outlined at the end of Chapter 3. The information and techniques to assess the probability of flooding on areas subject to alluvial fan flooding should be made available so that the floodplain managers and engineers involved in delineating alluvial fan flooding hazards can follow acceptable practices.

5. Expanding the technical and regulatory input base is necessary in order to successfully implement the NFIP in communities subject to alluvial fan flooding hazards. A mechanism is needed to provide FEMA with regular access to outside technical expertise in regard to the delineation and regulation process. For instance, EPA has a technical advisory board to provide input. Such a technical advisory group should include the representatives of scientists, engineers, local regulating bodies, and those being regulated. The purpose of a technical advisory group is not to regulate, but to help reduce exposure to floods by providing advice in regard to surveys of the hazard, methods for predicting flooding in certain settings, and similar scientific and technical issues. The committee was impressed with the knowledge base and the resources devoted to flood hazard management in the communities it visited during the course of this activity. Encouraging a regional approach to the review process will help FEMA to integrate and refine this knowledge base.

6. If FEMA elects to extend the current alluvial fan regulatory construct to any nonalluvial fan situation, it will need to change the term *alluvial fan flooding* to *uncertain flowpath flooding*. What you call something is important. The choice of the term *alluvial fan flooding* creates confusion and conflict when it is applied to nonalluvial fan areas. Effective management of flood hazards requires straightforward communication of risk using a vocabulary that is both technically sound and meaningful to the public. FEMA should set an example of effective risk communication by dealing directly with flood hazard uncertainty. Action to clarify confusion created by terminology will help transform the alluvial fan flooding concept from its current status as a vague, catch-all phrase to an administratively meaningful tool. But it should be noted that there will always be some confusion where the legal process demands precision in definition while natural systems reflect a continuum with gradations and great complexity.

Appendix A

Characteristics and Hazards Reported in Published and Unpublished Accounts of Alluvial Fan Flooding

Fan(s)/Location	Hazard(s)[a]	Reference	Comments
Carefree fan Carefree, Arizona	4,5,8,10	Hjalmarson, 1994[b]	Storm of October 6, 1993, produced runoff in all of the trenched distributary channels on the alluvial fan. Measurements and estimates of peak discharge show that a significant part of the peak discharge at the fan toe was from runoff on the developed soils of the fan itself. Distributary channels typically were filled to less than about 1/2 of bankfull capacity. Coarse sand deposited along channel beds by previous floods was remobilized and transported further toward and past the fan toe. The Carefree alluvial fan is on a piedmont, and the drainage basin heads on a pediment.
Chicago Creek Hazelton, British Columbia, Canada	1,3,4,5,7,9, 10	Gottesfeld, Mathewes, Gottesfeld, 1991	Major debris flow about 3580 BP covered about 300 ha with deposited debris. The flow was two to three orders of magnitude larger than other historic debris flows. This catastrophic event that formed part of the alluvial fan may be one of the oldest oral records of a major debris flow in North America. The Gitksan people of the area still talk about this debris flow. Major Holocene and historic flows are evident at the mouth of Chicago Creek, a tributary to the Skeena River, in northern British Columbia, Canada. Dating of catastrophic postglacial debris flow deposits indicates 5 major flows during the past 10,000 years, or a recurrence interval of about 2,000 years.
Cottonwood Canyon Bishop, California	1-5,7,9,10	Beaty, 1963	Rare account from eyewitness of debris and mudflows on July 26, 1952. Two hours after a heavy thunderstorm in the White Mountains to the east, a large flow of debris advanced down the alluvial fan. At and below the apex the flow was in a preexistent defined channel leading from the 9.5-km^2 basin. Debris spilled over channel walls and spread laterally to widths of 30 to 120 m. One large distributary channel of debris was formed by concentrated overflow on the outside of a gradual bend. The debris deposit was about 6.9 km long with a deposit of mud 1.1 km long followed by 20 to 25 cm of silt deposits for at least 0.8 km near the fan toe above Benton, Washington. Much of the deposited debris was remobilized by subsequent water flow during the event.
Cottonwood Creek Boise, Idaho	3,8,10	Waananen, Harris, Williams, 1971	Flooding of January 1965 caused local flooding and much sediment deposition in downtown Boise, Idaho. Estimated sediment yield from the 31.1-km^2 basin was about 38 metric tons per hectare.
Cottonwood Creek Boise, Idaho	3,8,10	Wyle, 1995[c]; Committee, 1995[d]	A wildfire burned much of vegetation in basin in 1959. Flood of August 20, 1959, deposited much sediment on fan. The unusual amount of sediment deposited by 1965 flood may be related to large amount of available sediment stored in basin tributaries

Fan(s)/Location	Hazard(s)[a]	Reference	Comments
			as a result of the wildfire in 1959. Considerably less sediment was deposited below the apexes of the Stuart and Crane Gulch alluvial fans a few miles to the west of the wildfire. The wildfire apparently destabilized the sediment production of an otherwise relatively stable (in an engineering context) drainage basin. The alluvial fans of Cottonwood Creek and Stuart and Crane Gulches are highly urbanized, and the paths of floodwater are significantly controlled by the conveyance capacity of the many streets that cross the fans at various angles relative to the general direction of fan slope.
Cucamonga Cucamonga, CA	3,4,6,8	Eckis, 1928	During the flood of February 16, 1927, small channels filled with debris and lessened the grade behind deposited debris, and spread floodwater to one or both sides of the debris dams. The spreading of floodwater and debris on the fans in the area, prior to the construction of the several flow control structures, is attributed to the many similar debris dams across channels.
	3,4,8	Singer and McGlone, 1971	Flood inundation maps for the flood of January 25, 1969, show a wide area of inundation below the apex above a flood control dam. Heavy rainfall from January 18 to 25 produced record flooding at Cucamonga Creek. The peak discharge from the 26.2-km^2 basin above the U.S. Geological Survey streamflow gage near Upland (number 11073470) was 399 km^3/s, or nearly 1.4 times the previous record flood of 1938. The peak discharge decreased downstream as floodwater was stored in the many percolation basins on the coarse-permeable material of the alluvial fan. Approximately 5.2 km^2 of fan area were inundated, mostly behind the percolation dams.
	8	DMA, 1985	Flow paths are confined by the high banks in the upper fan. Flow paths are relatively stable except for areas affected by structures developed after 1938.
		Committee, 1995[d]	Aerial photographs of August 3, 1989, reveal nearly all flow paths are affected by structures and floodwater is contained in structures. The compound fan has a large, deep trench in the old fan deposits on the south slopes of the San Gabriel Mountains. The region is geologically complex with marked differences among fans that form a *bajada* between the mountains and the Santa Ana River to the south. The Sierra Madre fault with a differential vertical movement of 1200 to 1500 m forms the contact between the rugged mountains and steep piedmont (Eckis, 1928, p. 226). The hydrographic apex of the younger Cucamonga Fan cannot be precisely defined

Fan(s)/Location	Hazard(s)[a]	Reference	Comments
			because many detention dam-type percolation basins have been constructed in the trench and fan and the mid and lower fan is extensively urbanized. The hydrographic apex may be from about 0.8 to about 2.4 km below the mountain front, where there is little channel incision. The old trench is about 27 m deep and 270 m wide near the mountain front, about 9 to 12 m deep and 340 m wide 900 m below where it expands to a width of 760 m with an average depth of about 1.5 m 2,100 m below the mountain front.
Day and Deer creeks Etiwanda, CA	3,4,5,	Waananen, 1969	Floodflows of January 25, 1969, discharged onto the alluvial fans, which coalesced and overflowed through residential areas. The peak discharge of 268 m^3/s at the U.S. Geological Survey streamflow gaging station on Day Creek near Etiwanda, California (no. 11067000) was greater than the 100-year flood.
	3,4	Singer and McGlone, 1971	Flood inundation maps show extensive flooding in the coalescing systems of distributary channels in the upper fan areas. Floodwaters of the two fans coalesced about 0.2 km below the mountain front. At the time of the flooding, there was a levee along the upper west side of the Deer Creek fan to divert flow to the east toward Day Creek. There also were two constructed flood channels on the western part of the Day Creek fan where the present feeder channel directs the flow from the 12.0-km^2 basin upslope. Floodflows of January 25, 1969, which were greater than the 100-year flood, exceeded the capacity of the levee and channels on both fans.
	3,4,8	Committee, 1995[d]	The flow paths of the several distributary channels on the fans apparently did not move during the flooding of 1969. The flow paths were examined using aerial photography on February 2, 1953, and August 3, 1989 (U.S. Natural Resources Conservation Service, Salt Lake City, Utah) and in the flood map report by Singer and McGlone (1971). The active appearing channels of both fans are interlacing over a width of 0.8 to 1.6 km within 3.2 km of the mountain front. Major channels of the two fans combine about 1.6 km below the mountain front. The pre- and post-flood flow paths are surprisingly similar, especially for such a large volume of floodwater and the reported tremendous quantities of transported sediment (Waananen, A.O., 1969, p. 14). The alteration of the natural drainage system by the levee and flood channels had some effect on the flow paths, but breaching occurred where major channels were intersected. With the exception on some minor effect of the structures, no change of flow paths over the fans was gleaned from the photographs. Clearly, even with the extensive and interlacing network of distributary channels, the

Fan(s)/Location	Hazard(s)[a]	Reference	Comments
	3,4,8	DMA, 1985	floodwater has followed the existing network of channels during the 36 years spanned by the photographs. Also concluded the flood channels in the area are rather stable. Five sets of aerial photographs taken over the period 1935 to 1969 were examined, and only one small new channel appeared to have formed on the Day Creek fan during the large flood of 1938. Except for the areas affected by the levees and channels constructed after the flood of March 2, 1938, no change in the channel pattern was detected.
Devil Canyon San Bernardino, California	4,8	DMA, 1985	Aerial photographs of distributary channels taken shortly before and after the second largest flood of record on March 2, 1938 show no movement of flow paths.
		Committee, 1995[d]	The flow paths on the alluvial fan are significantly altered by construction of recharge basins and a flood channel that conveys floodwater past the fan.
Furnace Creek Death Valley, California	1,2,4,5,9	Crippen, 1979	Flood hazard area estimated and mapped. Author mentions the "random" appearance of debris paths in upper very active portion of the fan.
	1,2,10	Miller, 1977	Account of July 1968 flood.
	1,3,4,5,9,10	Anstey, 1965	Flow of July 25, 1950, spread at apex and was lost to infiltration with deposited debris the result. Gorge near Furnace Creek Inn filled to a depth of 7.6 m. Gravel and rock rubble were deposited 3.2 km below apex.
	3,5,10	Hunt and Mabey, 1966	Probable flow of July 25, 1950, eroded about 30 percent of small 25-year-old earth embankments on fan. Also eroded about 15 percent to 29 percent of tributary wash crossings on 50-year-old trails in basin. Veneer of recent clay in places on upper fan. Flood boundaries are clearly defined by new erosion or deposition.
Glendora, California	1,3,8,10	Geisner and Price, 1971	Flood inundation maps show flooding of urban area on small fan. Flow paths were significantly influenced by street pattern that was about parallel and perpendicular to the fan axis.
Glenwood Springs Glenwood Springs, Colorado	3,4,9,10	Mock and Pawlak, 1983	Map of alluvial fans and debris flows of storm of July 24, 1977, which inundated about 0.8 km² of city with debris deposits. The city, which is located in western Colorado at 1,770-m elevation, is built on several alluvial fans that are active. Debris

Fan(s)/Location	Hazard(s)[a]	Reference	Comments
			flows are generated by soil slumps usually in colluvium and move at velocities from 0.3 to 12 m/s. The city experiences a debris flow about once every 4 years (since 1903). There are at least 32 streams that drain the steep surrounding slopes that have produced debris flows. The storm of July 24, 1977, produced debris flows from 20 tributary streams to the Roaring Fork River and the Colorado River.
Hanapah Canyon Death Valley, California	1,4,5,8	Hooke, 1995[e]	Personal eyewitness account of January 1969 flooding. Flow in many distributaries. Velocities of 5.5 and 7.7 m/s measured with the use of surface flouts in channel visually estimated to be 3.5 m wide and 0.4 m deep.
	4,5,8,9	Committee, 1995[d]	An excellent example of a fully trenched alluvial fan with shifting of the hydrographic apex far below the topographic apex. Large portions of the fan (the relict parts of the fan) are not subject to alluvial fan flooding. Numerous debris flow deposits with rounded lobes on the active portion of the fan.
Henderson Canyon Borrego Springs, CA	3,4,10	San Diego County, 1977; Committee, 1995[d]	Peak discharge on August 17, 1977, was 93.5 m³/s from the 16.3-km² basin and upper fan. The hydrographic apex of the active alluvial fan is located on the lower south side of the relict fan, and floodflow split into two channels below the apex. Flow fanned out and became sheetflow below the hydrographic apex, and fine sediment was deposited over a large area. The young active fan is on the right or south side of the relict fan that was formed by debris flow deposits. Incision of stable channels in the older boulder deposits has shifted the apex of active deposition.
Horseshoe Park Estes Park, Colorado	1,2,3,5,10	Tunbridge, 1983	An earth dam broke on July 15, 1982, in Rocky Mountain National Park, producing a 6- to 10-m-high wall of water and debris in the Roaring River. In just a few hours an alluvial fan covering 70 ha was formed in a valley at the mouth of the river 10 km below the dam. Much of the material that formed the fan was from incision in the glacial till along the river. The "instant" fan was up to 10 m thick.
	1-5,9,10	Blair, 1987	Peak water discharge at the fan was 340 m³/s based on dambreak model calibrated to eyewitness accounts. Released volume from dam of 830,000 m³. Eyewitness account at fan of leading edge of water, logs, and tree limbs followed by the flood peak 25 minutes later. Aerial photograph taken 5 hours after the flood arrived showed most of the fan had been formed. Most of fan building was from expanding sheetflooding that occurred in three phases. Some deposition was by supercritical flow on the flood

Fan(s)/Location	Hazard(s)[a]	Reference	Comments
			recession. Following the major discharge the flood deposit was modified by erosion and redeposition of the sheetflood deposits.
Howgill Fells Northwest England	3-10	Wells and Harvey, 1987	Storm of June 1982 destabilized hillslopes and produced both debris flows and water floods from a group of small basins. Thirteen fans with widely different physical characteristics were formed at tributary junctions. The new fans were up to 3 m thick and 100 m across. Fan processes were related to intrinsic fan-basin differences and to extrinsic characteristics such as sheep overgrazing by the Vikings in the 10th century. The 2.5-hour storm, with a recurrence interval of at least 100 years, produced extensive overland flow according to an eyewitness.
Lone Tree Creek Bishop, California	1-4,5,9,10	Beaty, 1963	See Cottonwood and Millner Canyons. The debris flow split near the mountain front forming an arm of debris deposits. Below the split the flow widened from 60 m to 150 m followed by mud and silt deposits similar to those at Cottonwood and Millner Canyons.
Lytle Creek San Bernardino, California	4,5,8	DMA, 1985	Largest two floods since 1920 were confined to defined channels between the mountain front and confluence with Cajon Creek. Based on comparison of aerial photographs taken shortly before and after floods of March 2, 1938, and January 25, 1969.
	5,8	McGlashan and Ebert, 1918	Bank erosion is suggested in account of bridge failure.
	4,8	Troxell, 1942	Bank erosion suggested during 1938 flood. Overbank flow near the Santa Anna River reported for great flood of 1861-62.
	8	Eckis, 1928	Description of channel below the apex suggests the channel is the same as present (1996) channel.
	1,4,5,8	Committee, 1995[d]	Based on comparison of aerial photographs of 1967 and 1989, flow paths are unchanged. The developed Soboba soils (NRCS, 1980) adjacent to the channel suggest the flow paths are stable. Channel capacity below the apex is about three times the magnitude of the largest flood of record since 1920 (U.S. Geological Survey gage no. 11062000).

Fan(s)/Location	Hazard(s)[a]	Reference	Comments
Magnesia Spring Canyon Rancho Mirage, California	1-5,7,10	FEMA, 1989; Committee, 1995[d]	Major flood of July 1979 breached the levee at the apex and flooded urban development in the city of Rancho Mirage. The estimated peak discharge of the flood was 170 m^3/s (Anderson-Nichols and Co.) from the 10.7-km^2 mountainous drainage basin, or nearly 3 times the magnitude of the 100-year flood (Thomas, Hjalmarson, and Waltemeyer, 1994). One death and $7 million in damage resulted. The effect of the breached levee on the distribution of flow in the several distributary channels is difficult to assess, but, based on comparison of aerial photographs for 1955 and 1982, the floodflow followed the pre-levee channel network, while some floodflow may have been directed from the right side of the fan to an existing large, incised channel on the left. Under natural conditions the flood hazard of this alluvial fan may be especially severe because the 2.6-km^2 fan area is relatively small for the drainage basin area. The peak discharge intensity, discharge per unit fan area, of the fan is fairly large, mostly because the fan is kept small as deposited material is periodically removed by Whitewater River, which forms the fan toe. A major flood and debris control structure has been constructed at the topographic apex.
Millner Canyon Bishop, California	1-5,9,10	Beaty, 1963	See Cottonwood Canyon. Large debris flow from the 26.2-km^2 basin deposited debris within and adjacent to the defined front where the flow split into three distributary channels. The debris overtopped the walls of the distributary channels for another 2 miles, covering a total width of about 0.8 km as observed by an eyewitness. Like the Cottonwood Canyon fan, there were mudflow deposits below the debris deposits followed by silt deposits.
	3,4,7,9,10	Beaty, 1970	The 1952 deposits and older deposits clearly described. The estimated recurrence interval of fan building events like the 1952 flood was 350 years.
	3,4,7,9,10	Beaty, 1989	Major debris flows can be expected at intervals of a few hundred years. The recurrence interval for Millner Canyon fan was 320 years based on C-14 dating.
Montrose Montrose, California	1-5,7,10	Chawner, 1935	Abnormally large volumes of sediment and floodflow were produced from a recently burned basin during storm of December 31, 1933, and January 1, 1934. A 4.6-m wall of water carrying houses, trees, boulders, and people was reported. The deposited sediment on the alluvial fan was equivalent to a 0.02-m layer from the 7.8-km^2 basin.

Fan(s)/Location	Hazard(s)[a]	Reference	Comments
Northumberland Canyon Austin, Nevada	3,4	DMA, 1985	Flow paths depicted on aerial photograph of June 19, 1981 suggest flood of August 7, 1979 of 217 m³/s was confined in channel 4.42 km below the mountain front where a young fan has formed. Channel patterns on the fan suggest flow initially spread in several distributary channels and eventually became sheetflow.
	1,3,4,7,10	Committee, 1995[d]	An approximately 0.8-km² area of recent deposits and a larger area of older oxidized deposits form the active alluvial fan, which is inset in the older deposits of this compound fan. The most recent deposits probably are from the 1979 flood that was about 10 times the magnitude of the 100-year flood for the 41.3-km² basin (Thomas, Hjalmarson, and Waltemeyer, 1994). The active alluvial fan is inset in the relict fan piedmont between the steep mountains to the east and salt marshes of a bolsom-like area to the west, which form much of the Big Smoky Valley. The large, deep trench in the relict deposits on the western slopes of the Toquima Mountain Range is incised for 4.4 km below the mountain front to the hydrographic apex. Trenching is restricted by a large bedrock outcrop 1.2 km below the mountain front. The cross section area of the trench is smaller but somewhat comparable in size to the trenches at the Cucamonga and Hanapah Canyon alluvial fans.
Picacho Peak Eloy, Arizona	4,6,9,10	Hjalmarson and Kemna, 1991	A small debris flow sometime between April and June 1989 followed the crest of alluvial fan in line with the channel axis at the fan apex. Deposited debris split 120 m below apex into two distinct lobes 60 and 45 m long. About 140 m³ of debris was deposited at average depths of 0.10 m with little, if any, water leaving the deposit bounds on the fan surface. The debris flow did not follow the axis of maximum slope below the topographic apex.
Saddle Mountain Arlington, AZ	1,3,4,6,8,10	USNRCS, 1987; Hjalmarson, 1995[d]	Major flood of September 2, 1984, filled and in places overflowed stable-trenched channels of relict fan. The peak discharge of the flood was 351 m³/s from the 22.3-km² upper relict fan and steep mountainous drainage basin. A conveyance-slope estimate on a trenched-tributary stream draining a small mountainous basin suggests the unit discharge of 15.7 m³/s/km² was nearly uniform over the mountains and upper piedmont, which are covered with varnished stones. Floodwater eroded several rills on the steep mountain slopes. Floodflow on the piedmont overtopped banks of stable distributary channels and inundated developed soils covered with varnished stones. A small amount of coarse material was deposited on varnished stones at one location where floodflow overtopped a relatively stable channel bank and spread over adjacent developed soils as sheetflow. No new distributary channels were formed. Floodwater

Fan(s)/Location	Hazard(s)[a]	Reference	Comments
			at nearly bankfull depths was in all defined distributary channels with sheetflow on much of the developed soils of the relict fan. There was some erosion damage to the side inlets and upslope banks of the Saddleback floodwater diversion channel located across the lower portion of the relict fan (U.S. Natural Resources Conservation Service, 1987). Much of the flood damage to the diversion structure was the result of inadequately sized floodwater inlets or because inlets were not located on some of the deceptively small distributary channels, which can convey shallow floodflow at near-critical velocities.
San Fernando Valley Los Angeles, California	3,4	King et al., 1981	Both confined flow and sheetflow are shown on maps of several riverine and alluvial fan areas flooded during 1934 through 1956.
Santa Monica Creek Carpinteria, California	3,4,8	Fenzel and Price, 1971	Flood of January 1969 overtopped stream banks at apex of alluvial fan and moved overland in thin sheets. Map of flooded area shows that most of flooding was in an agricultural area.
Sydney fans Australia	3–8,9	Scott and Erskine, 1994	Storm of February 2–4, 1990, caused a variety of responses on 12 small fans including avulsions, progressive aggradation, localized erosion, and fanhead trench reworking. One fan experienced no detectable change, while 7 with trenches experienced trench reworking. Three fans had localized deposition, 2 had spatially disjunct erosion and deposition and/or channel avulsions. Avulsion was related to channel filling with a steeper alternate path leading from the newly formed topographic high to a topographic low. Although the 12 fans are very small and formed by water flood processes, a potentially significant result identified by the authors was a threshold slope for fanhead trench initiation.
Wadi Mikeimin Sinai Desert	1,3,7,9,10	Schick and Lekach, 1989	An alluvial fan was formed during the flood of January 1971 at the confluence of Wadi Mikeimin and Wadi Watir. The fan dammed Wadi Watir for 21 months before it was washed out by a major flood. The peak of 68.5 m^3/s from the 13-km^3 basin deposited 6,200 m^3 of coarse predominantly stratified sediment. The total flow of the flood was in excess of 100,000 m^3. The slope of the channel at the fan apex was 0.087.

Fan(s)/Location	Hazard(s)[a]	Reference	Comments
Wasatch Front Salt Lake City, Utah	2,7,9,10	Patton and Baker, 1976	Between 1847 and 1938, over 500 cloudburst floods were recorded along the Wasatch mountain front with reports of damage. Although not specifically given, many, if not most, of these floods were on alluvial fans because of the great concentration of fans along the front. Between 1939 and 1969, accounts of 836 flash floods were reported, the majority of which were along the front. Flash floods commonly cause debris flows, which greatly increase property damage.
	9,10	Mathewson and Keaton, 1990	Based on historic records of flood damage, the damaging debris flows during the springs of 1983 and 1984 may have a recurrence interval of 100 years if the type of initiating mechanism is considered irrelevant. However, only 3 to 5 prehistory debris flows have occurred during the past 12,000 years. Many large debris flows occurred in the 1920s and 1930s that were triggered by thunderstorms. At that time the slopes had been cleared of dense vegetation by burning to encourage new growth of grasses and improve grazing. The debris flows of the 1920s and 1930s were totally different from those of 1983 and 1984. Recurrence intervals (RI) of Davis County debris flows along the Wasatch mountain front are difficult to define as shown by the following: RI = 100 years based on historic record, RI = 3,000 years based on stratigraphic record, RI = 10,000 years based on phenomena causing the 1983 and 1984 flows.
White Tanks Phoenix, Arizona	1,3,4,5,7,10	Hjalmarson and Kemna, 1991	Active alluvial fan with evidence of recent channel movement within active fan boundaries probably as a result of frequent floods. Hydrologic apex is below trenched channel in relict deposits where there was no evidence of flow path movement. Also, no evidence of recent channel movement in lower portion of active alluvial fan where there has been little recent floodflow because floodwater of the common floods that passes the hydrographic apex is lost to infiltration.
	1,3,4,5,7,10	CH2M HILL, 1992	Aerial photographs for 1942 through 1979 comparison shows that about 24 percent of the 35 km of channels shifted location or were formed. Paleoflood analysis indicates no major floods during this period.
Wild Burro Tucson, Arizona	3,4,5,8	Pearthree et al., 1995	Flood 1988 on the southern piedmont of the Tortolita Mountains followed distinct and existing flow paths separated by large dry areas. Floodflow of the rare flood ($p <$ 0.01) split into 42 paths. Dry areas separating the flow were more than one half of the alluvial fan. Surficial geology and flood boundaries were mapped to produce a rare display of alluvial fan flooding on a system of trenched distributary channels.

[a]Reported flood characteristics and hazards:
1. High flow velocities
2. Flash floods and possibly translatory waves
3. Sheetflow and/or sheetflooding
4. Distributary flow
5. Unstable channel boundaries including bed and bank erosion and remobilization of deposited sediment
6. Stable channel boundaries
7. Movement of flow paths
8. Stable flow paths
9. Debris flow (including hyperconcentrated and mud flows)
10. Alluviation

[b]Based on unpublished data and observations of H.W. Hjalmarson on file at the Flood Control District of Maricopa County, Arizona.
[c]Based on oral communication with Jim Wyle of City of Boise Idaho Public Works, April 14, 1995.
[d]Based largely on unpublished field observations made during the course of this study by one or more members of the Committee on Alluvial Fan Flooding.
[e]From unpublished observations and measurements in field notes for flood of January 1969.

REFERENCES

Anderson-Nichols and Co. 1981. Floodplain Management Tools for Alluvial Fans. Federal Emergency Management Agency Contract EMW-C-0715.

Anstey, R. L. 1965. Physical Characteristics of Alluvial Fans: United States Army Natick Laboratory, Technical Report ES-20.

Beaty, C. B. 1963. Origin of alluvial fans, White Mountains, California and Nevada. Pp. 66-73 in Modern and Ancient Alluvial Fan Deposits, T. H. Nilsen, ed. New York: Van Nostrand Reinhold Co.

Beaty, C. B. 1970. Age and estimated rate of accumulation of an alluvial fan, White Mountains, California, U.S.A. American Journal of Science 268:50-77.

Beaty, C. B. 1989. Great big boulders I have known. Geology 17:349-352.

Blair, T. C. 1987. Sedimentary processes, vertical stratification sequences, and geomorphology of the Roaring River alluvial fan, Rocky Mountain National Park, Colorado. Journal of Sedimentary Petrology 57(1):1-18.

Chawner, W. D. 1935. Alluvial fan flooding-the Montrose, California, flood of 1934: Geological Review, American Geographical Society of New York 25(2):255-263.

CH2M Hill. 1992. Alluvial fan data collection and monitoring study: On file at Flood Control District of Maricopa County, Arizona.

Crippen, J. R. 1979. Potential hazards from floodflows and debris movement in the Furnace Creek area, Death Valley National Monument, California-Nevada: U. S. Geological Survey Open-File Report 79-991. Reston, Va.: U.S. Geological Survey.

DMA Consulting Engineers. 1985. Alluvial Fan Flooding Methodology—An Analysis. Federal Emergency Management Agency Contract EMW-84-C-1488. Washington, D.C.: FEMA.

Eckis, R. 1928. Alluvial fans of the Cucamonga district, Southern California: Journal of Geology 36:224-247.

Federal Emergency Management Agency (FEMA). 1989. Alluvial Fans: Hazards and Management. FEMA Document 165. Washington, D.C.: FEMA.

Hjalmarson, H. W., and S. P. Kemna. 1992. Flood Hazards of Distributary-Flow Areas in Southwestern Arizona. U.S. Geological Survey Water Resources Investigations Report 91-4171. Reston, Va.: U.S. Geological Survey.

Hunt, C. B., and D. R. Mabey. 1966. Stratigraphy and Structure, Death Valley, California. U. S. Geological Survey Professional Paper 494-A. Reston, Va.: U.S. Geological Survey.

King, E. J., J. C. Tinsley, and R. F. Preston. 1981. Maps Showing Historic Flooding in the San Feranado Valley, California, 1935 to 1956. U. S. Geological Survey Open-File Report 81-153. Reston, Va.: U.S. Geological Survey.

Mathewson, C. C., and J. R. Keaton. 1990. Multiple Phenomena of Debris-Flow Processes: A Challenge for Hazard Assessments. Pp 549-554 in Proceedings of the International Symposium, Hydraulics/Hydrology of Arid Lands: Hydraulics Division of American Society of Civil Engineers. New York: ASCE.

McGlashan, H. D., and F. C. Ebert. 1918. Southern California Floods of January 1916. U.S. Geological Survey Water-Supply Paper 426. Reston, Va.: U.S. Geological Survey.

Miller, G. A., 1977, Apprasial of the Water Resources of Death Valley, California-Nevada. U. S. Geological Survey Open-File Report 77-728. Reston, Va.: U.S. Geological Survey.

Mock, R. G., and S. L. Pawlak. 1983. Alluvial fan hazards at Glenwood Springs. Pp. 221-233 in Special Publication on Geological Environment and Soil Properties, R. N. Yong, ed. New York: American Society of Civil Engineers.

Natural Resources Conservation Service (NRCS). 1987. Engineering report of Saddleback Diversion, Harquahala Valley watershed, Maricopa County, Arizona. U. S. Department of Agriculture.

Natural Resources Conservation Service (NRCS). 1980. Soil Survey of San Bernardino County—Southwestern Part, California. U. S. Department of Agriculture.

Natural Resources Conservation Service (NRCS). 1980. Soil Survey of Riverside County, California, Coachella Valley Area. U. S. Department of Agriculture.

Natural Resources Conservation Service (NRCS). 1973. Soil Survey of San Diego County, California: U. S. Department of Agriculture.

Patton, P. C., and V. R. Baker. 1976. Morphology and floods in small drainage basins subject to diverse hydrogeomorphic controls: Water Resources Research 12(5):941-952.

Pearthree, P. A., P. K. House, and K. R. Vincent. 1995. Detailed Reconstruction of an Extreme Alluvial Fan Flood on the Tortolita Piedmont, Pima County, Arizona. Arizona Geological Survey Open-File Report. Tucson: Arizona Geological Survey.

San Diego County. 1977. Storm Report for Tropical Storm Doreen, August 15-17, 1977. County of San Diego Department of Sanitation and Flood Control.

Schick, A. P. 1995. Fluvial processes on an urbanizing alluvial fan: Eilat, Israel: American Geophysical Union Geophysical Monograph 89:209-218.

Scott, P. F., and W. D. Erskine. 1994. Geomorphic effect of a large flood on fluvial fans. Earth Surface Processes and Landforms 9:95-108.

Singer, J. A., and P. McGlone. 1971. Flood of January 1969 Near Cucamonga, California. U. S. Geological Survey Hydrologic Atlas HA-425. Reston, Va.: U.S. Geological Survey.

Thomas, B. E., H. W. Hjalmarson, and S. D. Waltemeyer. 1994. Methods for Estimating Magnitude and Frequency of Floods in the Southwestern United States. U.S.Geological Survey Open-File Report 93-419. Reston, Va.: U.S. Geological Survey.

Troxell, H. P. 1942. Floods of March 1938 in Southern California. U.S. Geological Survey Water-Supply Paper 844. Reston, Va.: U.S. Geological Survey.

Tunbridge, I. P. Alluvial fan sedimentation of the Horseshoe Park Flood, Colorado, U. S. A., July 15th, 1982. Sedimentary Geology 36:15-23.

U. S. Army Corps of Engineers. 1993. Assessment of Structural Flood-Control Measures on Alluvial Fans. Davis, Calif.: USACE Hydrologic Engineering Center.

Waanamen, A. O., D. D. Harris, and R. C. Williams. 1971. Floods of December 1964 and January 1965 in the Far Western States, Part 1, Description. U.S. Geological Survey Water-Supply Paper 1866-A. Reston, Va.: U.S. Geological Survey.

Wells, S. G., and A. M. Harvey. 1987. Sedimentologic and geomorphologic variations in storm-generated alluvial fans, Howgill Fells, northwest England. Geological Society of America Bulletin 98:182-198.

Appendix B

Sources of Data

Source Agency	**Extent of Data/Information Content**

1. U.S. Bureau of Land Management

Western United States:
National Applied Resource Science Center
DFC, Bldg. 50 RS-120
P.O. Box 25047
Denver, CO 80225-0047
Tel. (303) 236-7991
FAX (303) 236-7990

For information about coverage in Alaska contact:
Alaska State Office
222 W. 7th Avenue, #13
Anchorage, AK 99513-7599
Tel. (907) 271-5063

The following types of photography are available:

Resource photography: Flown over large area generally at scales of 1:12,000 or 1:24,000. Film may be black-and-white, natural color, or false color-infrared.

Riparian photography: Flown over small stream segments generally at scales of 1:2,400, 1:4,800, or 1:6,000. Primarily false color-infrared film.

Photogrammetric photography: Generally flown at scales ranging from 1:2,400 to 1:12,000 for photogrammetric applications.

2. U.S. National Park Service (NPS)

Maps may be ordered by writing to the superintendent of each National Park Service unit. For more information, contact the Regional Office in your area of interest or the NPS Office of Public Inquiries, Room 1013, Washington, D.C., 2024, or NPS, Denver Service Center, 655 Parfet Street, P.O. Box 25287, Denver, CO, 80225.

NPS maps use United States Geological Survey (USGS) topographic maps as a base for planimetrically accurate data. The parks then supplement the base with up-to-date information—new road and trail alignments, building locations, etc. The information is focused towards the needs of park visitors, including road access, recreation, major topographic features, administrative boundaries, and other points of interest. Each park has its own focus—recreation, historic, ecosystems, general information, or interpretation.

Many park maps have shaded relief art, an artist's rendering of topographic features derived from USGS contour lines on the base map.

Scales and revision cycles vary from map to map.

The NPS also contracts for aerial photography over and adjacent to U.S. national park lands and other areas such as national monuments and national recreational areas. They are available through the EROS Data Center in Sioux Falls.

3. U.S. Geological Survey (USGS)

Aerial Photographs and Satellite Photographs

Earth Science Information Center
U.S. Geological Survey
507 National Center

Multiple aerial photographic coverage of the United States; satellite hand-held photos along paths in the United States and other parts of the world. Indexes

Source Agency	**Extent of Data/Information Content**

Reston, VA 22092
Tel. (703) 648-6045
FAX (703) 648-5548
Toll-free number 1-800-USA-MAPS

are available for NAPP, NHAP, and satellite hand-held photographs. The Earth Science Information Center (ESIC) Aerial Photography Summary Record System (APSRS) provides data on current and photo coverage of the United States.

 Sioux Falls ESIC
U.S. Geological Survey
EROS Data Center
Sioux Falls, SD 57198
Tel. (605) 594-6151
FAX (605) 594-6589

Aerial photos at various scales, with different emulsions, and information on camera characteristics, cloud cover and sources for obtaining copies. Hand-held satellite photos in black-and-white and color.

Flood-Prone Area Maps

Flood-prone area maps, although not a published series, are available, by quadrangle name, from the USGS Water Resources Division District Office in the state of interest.

Locations of the USGS Water Resources Division district offices can be obtained by contacting

U.S. Geological Survey
Hydrologic Information Unit
419 National Center
Reston, VA 22902
Tel. (703) 648-6817

Flood-prone areas are outlined on standard USGS topographic quadrangles at a scale of 1:24,000 as part of the national program for managing flood losses in urban areas mandated by the National Flood Insurance Act of 1968 and the recommendations of the Task Force on Federal Flood Control Policy (89th Congress). Efforts to produce these maps began in 1969. As of Fiscal Year 1976, over 12,000 quadrangles had been mapped.

Orthophotoquads

Earth Science Information Center
U.S. Geological Survey
507 National Center
Reston, VA 22092
Tel. (703) 648-6045
FAX (703) 648-5548
Toll-free number 1-800-USA-MAPS

Orthophotoquads contain no symbolized features, and a minimal number of names. Because they show a wealth of planimetric details and land use and land cover information that is not symbolized on conventional line maps, they make excellent supplements to the published maps. Pilot projects have demonstrated that digital scanning of aerial photos and rectification of scanned imagery to horizontal and vertical ground control can produce orthophoto digital data bases from which orthophotoquads at 1:12,000 and 1:24,000 are generated.

Quadrangle Format Maps

Earth Science Information Center
U.S. Geological Survey
507 National Center
Reston, VA 22092
Tel. (703) 648-5920
FAX (703) 648-5548

7.5-minute map series: conterminous United States, Hawaii, and territories at 1:24,000 or 1:25,000; Puerto Rico at 1:20,000.

15-minute map series: Alaska at 1:63,360.

Source Agency	**Extent of Data/Information Content**
Toll-free number 1-800-USA-MAPS	30- by 60-minute map series: conterminous United States and Hawaii at 1:100,000.
U.S. Geological Survey Map Distribution Federal Center, Box 25286 Denver, CO 80225 Tel. (303) 202-4700	Quadrangle format maps are a multicolored topographic base map series of the U.S. Geological Survey (USGS). Most USGS topographic maps use contours to show the shape and elevation of the terrain. Elevations are usually shown in feet, but on some maps they are in meters. Contour intervals vary, depending mainly on the scale of the map and the type of terrain. The maps show and name prominent natural and cultural (manmade) features. Those at scales of 1:24,000 show an area in detail. Such detail is useful for engineering, local area planning. Less detail is shown at scales of 1:63,360 and 1:100,000. They cover land areas and are used in land management and planning. Maps at a scale of 1:250,000 cover very large areas on each sheet and are used in regional planning.
Quaternary Geologic Atlas of the United States	
Earth Science Information Center U.S. Geological Survey 507 National Center Reston, VA 22092 Tel. (703) 648-6045 FAX (540) 648-5548 Toll-free number 1-800-USA-MAPS	Multicolored maps on topographic bases in 4- by 6-degree quadrangle units; scale 1:1 million; show regional distribution of Quaternary (surficial) geologic materials in the conterminous United States and adjoining areas.
U.S. Geological Survey Information Services Bldg. 810 Denver Federal Center, Box 25286 Denver, CO 80225 Tel. (303) 202-4700	Maps are available for 18 of the 53 proposed quadrangles in the atlas, and another 12 maps have been approved for publication and are currently in production. The rest are being compiled. Most of the maps available or in production are in the eastern and central United States. For availability, see sources listed under Miscellaneous Investigations Series (I Series) maps.
Water Data	
Information about water data is available on request from	Water quantity and quality data for geographic regions of the United States are available in print form and are machine-readable files.
The Assistant Chief Hydrologist for Operations 441 National Center Reston, VA 22092 Tel. (703) 648-5305	Data are available for all parts of the nation where the USGS has a data collection program. Availability can vary by state, according to the program of the USGS District Office for that state.
Water data questions may also be directed to the National Water Information Clearinghouse (NWIC): 1-800-426-9000.	

Source Agency	**Extent of Data/Information Content**

General information on water data and referrals to field offices may be obtained from

U.S. Geological Survey
Hydrologic Information Unit
419 National Center
Reston, VA 22092
Tel. (703) 648-6817

Or contact the District Chief in the state of interest to obtain data.

Water Data Reports may be purchased from:

The National Technical Information Service
5285 Port Royal Road
Springfield, VA 22161
Tel. (703) 487-4763

Bulletins, Professional Papers, Water Supply Papers and Other Book Series

Earth Science Information Center
U.S. Geological Survey
507 National Center
Reston, VA 22092
Tel. (703) 648-6045
FAX (703) 648-5548
Toll-free number 1-800-USA-MAPS

U.S. Geological Survey
Information Services
Denver Federal Center, Box 25286
Denver, CO 80225
Tel. (303) 202-4200

Bulletins: Reports on the results of resource studies and of geologic and topographic investigations, as well as collections of short papers related to a specific topic; generally more limited in scope or geographic coverage than professional papers. Some contain geologic or geophysical maps.

Professional papers: Comprehensive reports on the results of resource studies and of geologic, hydrologic, and topographic investigations. Also include collections of related papers addressing different aspects of a single scientific topic. Many contain geologic or geophysical maps.

Other: The USGS publishes a wide variety of reports containing the results of water resources investigations, many of which depict maps of the areas studied.

Also available are catalogues of USGS publications. A monthly list of new USGS publications is available free from

U.S. Geological Survey
582 National Center
Reston, VA 22092

Random products available are catalogued in the following sources:

"Publications of the Geological Survey, 1879-1961," "Publications of the Geological Survey, 1962-1970," "Publications of the Geological Survey, 1971-1981," and annual supplements for 1982 and later years. New bulletins are listed in the monthly catalog, "New Publications of the U.S. Geological Survey."

Source Agency	**Extent of Data/Information Content**

In addition, geologic maps published as plates or figures in bulletins are catalogued in the following sources:

• GEOINDEX, a computerized data base produced by the U.S. Geological Survey (USGS) that contains bibliographic information for published geographic maps.

• USGS Geologic Map Indexes (GMIs) for the 50 states and several territories.

These sources are available at Earth Science Information Center offices, in USGS libraries, and may be available at other earth science libraries, such as those of universities.

4. USDA Farm Service Agency (FSA)

USDA FSA Aerial Photography Field Office
Sales Branch
2222 West 2300 South
P.O. Box 30010
Salt Lake City, UT 84130-0010
Tel. (801) 975-3503

Aerial photography is available for the conterminous United States, Hawaii, and portions of Alaska. Contact the Sales Branch for details or for a comprehensive listing of available coverage.

Aerial photography is available from various types of film (black-and-white, natural color, color-infrared) at scales from 1:6,000 up to 1:120,000. The precise visual information provided by aerial photography can be used in many ways. The myriad of applications include conservation practices, locating field boundaries, tax assessment, urban development, pollution studies, and watershed studies.

5. U.S. Forest Service

Reports and Wilderness and Special Area Maps

Regional Offices:
Northern Region
Federal Building
200 East Broadway Street
P.O. Box 7669
Missoula, MT 59807
Tel. (406) 329-3511

Intermountain Region
Information Center
2501 Wall Avenue

The Special Area Maps coincide with USGS 7.5-minute quadrangle maps but may conform to other geographic lines or to topographic or cultural features. Because of the wide variety of needs and interest found within special areas, there is no specific standard scale for the wilderness and special area maps.

The USFS contracts for and acquires aerial photographic coverage over national forest lands and other areas related to their mandates. A variety of scales and film types have been employed. Most of these photographs are available through the USDA-

Source Agency	Extent of Data/Information Content
Union Station Ogden, UT 84401 Tel. (801) 625-5306	ASCS Aerial Photography Field Office in Salt Lake City.
Southern Region Information Center Suite 154 1720 Peachtree Road, NW Atlanta, GA 30309 Tel. (404) 347-2384	The USFS has nine regions with regional headquarters where regional aerial photographic coverage can often be viewed and ordered. Consult the government section of your telephone book for details.
Rocky Mountain Region 740 Sims Avenue P.O. Box 25127 Lakewood, CO 80225 Tel. (303) 275-5350	Inquiries may also be referred to Division of Engineering, U.S. Forest Service, Washington, D.C., 20250.
Pacific Southwest Region 630 Sansome Street San Francisco, CA 94111 Tel. (415) 705-2874	
Eastern Region 310 West Wisconsin Avenue Room 500 Milwaukee, WI 53203 Tel. (414) 297-3290	
Southwestern Region Public Affairs Office Federal Building 517 Gold Avenue SW Albuquerque, NM 87102 Tel. (505) 842-3292	
Pacific Northwest Region 333 SW First Street Portland, OR 97204 Tel. (503) 666-0771	
Alaska Region Public Services Section Federal Office Building 709 West Ninth Street P.O. Box 21628 Juneau, AK 99802-1628 Tel. (907) 586-8806	

Source Agency	Extent of Data/Information Content

6. U.S. Natural Resources Conservation Service

Soil Surveys and Soil Survey Geographic Database (SSURGO)

To obtain SSURGO soil spatial and attribute data, contact

National Cartographic and Geospatial Center
USDA–Natural Resources Conservation Service
P.O. Box 6567
Fort Worth, TX 76115
Tel. (817) 334-5559
FAX (817) 334-5469
Toll-free number 1-800-672-5559

SSURGO data are available for selected counties and areas throughout the United States and its territories. A soil survey digitizing status map and list of surveys digitized are available.

To obtain soil surveys and technical information about the use of soils data, please contact a NRCS state soil scientist in your state or contact

National Soil Survey Center
USDA–Natural Resources Conservation Service
Federal Building, Room 152
100 Centennial Mall, North
Lincoln, NE 68508
Tel. (402) 437-5423

Digitizing is done by line segment (vector) format in accordance with NRCS digitizing standards and specifications. The mapping bases used meet national map accuracy standards and are either orthophotoquads or 7.5-minute quadrangles. SSURGO data are collected and archived in 7.5-minute quadrangle units, and distributed as complete coverage for a county or area usually consisting of 10 or more quad units. Soil boundaries ending at quad neatlines are joined by computer to adjoining orthophotoquad maps to achieve an exact match.

SSURGO is linked to a Soil Interpretations Record, an attribute relational data base, which gives the proportionate extent of the component soils and their properties for each map unit. SSURGO contains one to three components. The Soil Interpretations Record data base includes over 25 soil, physical, and chemical properties for approximately 18,000 soil series recognized in the United States. Information on soil survey reports that can be queried from the data base includes available water capacity, soil reaction, salinity, flooding, water table, bedrock, and interpretation for septic tank limitations, engineering, cropland, woodland, and recreation development.

7. National Ocean Service (NOS)

NOAA, National Ocean Service
HSB, Data Control Section, N/CG243
WSC1, Room 404
6001 Executive Boulevard
Rockville, MD 20852
Tel. (301) 443-8408

Topographic maps represent a unique and comprehensive record of the coastline and the adjacent waters, showing conditions existing on particular dates and a record of the changes that have occurred from both natural and artificial causes. Most of the shoreline maps have been compiled at scales of 1:10,000 or 1:20,000. A number of harbor areas have been completed at 1:5,000 scale.

Topographic and photogrammetric surveys (shoreline maps) are surveys of the land features of an area. As their extent is limited by varying nautical chart requirements, they generally cover a distance of 1 to 10 kilometers inland from the shoreline. Topographic surveys vary not only in coverage but also in content.

Source Agency	**Extent of Data/Information Content**

8. U.S. Army Corps of Engineers (USACE)

USACE Publications Department
2803 52nd Avenue
Hyattsville, MD 20781
Tel. (301) 394-0081

USACE Publication Distribution Center
2800 Eastern Boulevard
Baltimore, MD 21220-2896
Tel. (410) 682-8524

Technical information on the Corps of Engineers' involvement with specific alluvial fans can be obtained by contacting individual district offices where that fan is located. General information on Corps policy programs or research can be obtained by contacting the distribution center.

9. Federal Emergency Management Agency (FEMA)

For information on the digital products, contact

Assistant Administrator
Federal Insurance Administration
Office of Risk Assessment
500 C Street, SW
Washington, DC 20472

For information on the analog products, contact

Service Center
P.O. Box 1038
Jessup, MD 20794-1038
Tel. 1-800-358-9616

Digital thematic overlay and analog maps of flood hazards. These products map delineating areas expected to be inundated by the 1 percent chance per year flood event, base flood elevation, floodway, and other flood risk data.

10. Tennessee Valley Authority (TVA)

Aerial photographs, topographic maps, orthophotoquads, and reports

TVA Map Sales
HB1A
1101 Market Street
Chattanooga, TN 37402-2801
Tel. (423) 751-MAPS (6277)

Black-and-white panchromatic, color and color-infrared aerial photographs of areas within the Tennessee Valley Region at various dates from 1933 to the present time.

National series, large-scale base cartographic data in cooperative mapping program administered by the United States Geological Survey. This series is at 1:24,000 scale. The Tennessee Valley Authority (TVA) produces, revises, and maintains 805 7½-minute quadrangles in this series.

This series of orthophotoquads is currently being developed. The orthophotoquads will be produced at 1:12,000 scale, in 3¾-minute quarter quadrangle format.

Source Agency	**Extent of Data/Information Content**
11. National Archives and Record Service (NARS)	
Reference Services Staff of the National Archives and Records Administration may be contacted at National Archives and Records Service Cartographic Branch Rm 3320 8601 Adelphi Road College Park, MD 20740-6001 Tel. (301) 713-7030	The NARS archives historical federal geographic data products including a variety of photo and map data and, in particular, aerial photographs flown by the government previous to 1942. Generally, one or two dates of aerial photography coverage are available for counties in the United States. They publish a pamphlet that describes available coverage and provides ordering details.
12. Library of Congress	
The division can be reached at (202) 707-MAPS.	The Map and Geography Division of the Library of Congress has an extensive collection of current and historical geographic data products from federal and nonfederal sources.
13. State geologists or geological surveys	
Contact state geologist in your state.	Several state geological information agencies have maps and reports of geologic features.
14. State floodplain management agencies	
Contact state agency.	A variety of geographic data products are available for inspection at some state agencies. Most state agencies have water and flood data and information on federal and state floodplain management. Some agencies have large-scale topography for limited areas.
15. University Libraries and academic institutions	
Contact your state university.	A wide variety of geographic data products and reports. Many universities have an extensive collection of aerial photographs, topographic maps, geologic maps and other maps of other geographic information.
16. County floodplain management agencies	
Contact your local agency.	Often the best source of detailed topography, flood data, and flood maps. A variety of geographic data products area available for inspection at some county agencies
17. Long-time residents	Can be good source of historical information on the frequency and severity of flooding problems.

Source Agency	Extent of Data/Information Content
18. Newspapers	Often a good source of historical information.
19. Technical journals	A wide variety of information related to flooding.
20. University theses	A wide variety of information related to alluvial fan flooding. The specific location and availability will vary, but would include libraries of colleges or departments of agriculture, water resources, geography, etc.

Appendix C

Biographical Sketches of Committee Members

STANLEY A. SCHUMM, *chair*, is University Distinguished professor, in the Department of Earth Resources, at Colorado State University and a consultant at Ayres Associates. Dr. Schumm is an expert in geomorphology and fluvial dynamics. He also served 13 years as a research geologist for USGS in the Water Resources Division. He has served on numerous NRC committees, including the Committee on Remote Sensing (CORSPERS) Geology Panel—1973-77, the Committee on Disposal of Excess Spoil—1980-81, the Committee on Hydraulic Models—1981-82, and the Committee on Solid-Earth Sciences (one of the panel chairs as well)—until 1992. Dr. Schumm has a B.A. from Upsala College, and a Ph.D. in geomorphology from Columbia University.

VICTOR R. BAKER is Regents Professor and Head of the Department of Hydrology and Water Resources at the University of Arizona. He is also professor of geosciences and planetary sciences at the University of Arizona. His research interests include geomorphology; fluvial geomorphic studies in the western United States, Australia, India, Israel, and South America; flood geomorphology; paleohydrology; Quaternary geology; natural hazards; geology of Mars and Venus; and philosophy of earth and planetary sciences. He has spent time as a geophysicist for USGS and as an urban geologist. He has served on various committees and panels of the National Research Council, including the Panel on Global Surficial Geofluxes, the Global Change Committee Working Group on Solid Earth Processes, and the Panel on Scientific Responsibility and Conduct of Research. He is currently chair of the U.S. National Committee for International Union for Quaternary Research, a committee of the National Research Council's Board on Earth Sciences and Resources. Dr. Baker holds a B.S. from Rensselaer Polytechnic Institute, and a Ph.D. from the University of Colorado.

MARGARET (PEGGY) F. BOWKER is president, owner, and principal engineer for Nimbus Engineers. An expert in surface water hydrology and hydraulics, and flood control and floodplain management for arid and semiarid climates, she has been actively involved in observing, delineating, and regulating alluvial fans for the past 14 years. She worked in floodplain management for Pima County Arizona Flood Control District in the early 1980s, and then focused on disaster assistance for FEMA. Nimbus has performed several FEMA flood insurance studies in

Nevada alluvial fan communities and has conducted a survey of communities subject to alluvial fan flood hazards for FEMA through the Association of State Flood Plain Managers. She has a B.S. in civil engineering from the University of Nevada and is a licensed professional engineer in Arizona and Nevada.

JOSEPH R. DIXON is a supervisory civil engineer and section chief in the Water Resources Branch of the Planning Division for the U.S. Army Corps of Engineers. His primary expertise is in flood control planning and the preparation of feasibility reports for Congress. He has recently worked on federal flood control studies on alluvial fans in Nevada and Arizona. He is involved with the implementation of Corps policy using a risk and uncertainty approach to evaluate the federal economic interest in participating in flood control projects. He holds a B.S. from the University of Arizona and an M.S. in civil engineering (sanitary) from California State University, Long Beach.

THOMAS DUNNE is a professor in the School of Environmental Science and Management at the University of California at Santa Barbara. He is a hydrologist and a geomorphologist, with research interests that include alluvial processes; field and theoretical studies of drainage basin and hillslope evolution; sediment transport and floodplain sedimentation; debris flows and sediment budgets of drainage basins. He served as a member of the WSTB Committee on Water Resources Research and Committee on Opportunities in the Hydrologic Sciences and was elected to the National Academy of Sciences in 1988. Dr. Dunne holds a B.A. from Cambridge University and a Ph.D. in geography from the Johns Hopkins University.

DOUGLAS HAMILTON is a professional engineer and an independent hydrologic consultant with experience in a range of water resources issues including environmental impact analysis, computer modeling, and mitigation of natural hazards. He has assisted several communities with the evaluation of alluvial fan flood maps under the National Flood Insurance Program. His practice serves both public and private clients. He has worked at the Hydrologic Engineering Center, and is an instructor for University of California, Davis, Extension. He has a B.S. from Harvey Mudd College and an M.S. from the University of California, Davis.

HJALMAR W. HJALMARSON is a consulting hydrologist and registered professional civil engineer with a long interest in river hydraulics and arid land hydrology. He is retired from the Water Resources Division of the U.S. Geological Survey, where he conducted studies and published reports on water resources, surface water hydrology, the magnitude and frequency of floods in the southwestern United States, stochastic hydrology, and alluvial fan and riverine flooding. He has worked with the National Flood Insurance Program as a technical specialist since 1970. He also taught at the University of Arizona as an adjunct professor. He currently is studying flood characteristics of piedmont plains in central Arizona as a consultant for the Flood Control District of Maricopa County. He has a B.S. in engineering from Arizona State University.

DOROTHY MERRITTS is associate professor of geosciences at Franklin and Marshall College. Her research interests include alluvial fan processes in California, geomorphology, tectonics, hydrology, and soils. She is co-author of a textbook on environmental geosciences, with emphasis

emphasis on surficial processes and surface water hydrology. She serves on the Committee on Undergraduate Science Education (a standing NRC Commission on Life Sciences committee). She received a B.S. from Indiana University of Pennsylvania, an M.S. from Stanford University, and a Ph.D. in geology from the University of Arizona.

Appendix D

Glossary and List of Acronyms

Alluvial   Pertaining to or composed of alluvium, or deposited by a stream or running water.

Alluvium   A general term for clay, silt, sand, gravel, or similar unconsolidated detrital material deposited during comparatively recent geologic time by a stream or other body of running water as a sorted or semisorted sediment in the bed of the stream or its floodplain or delta, or as a cone or fan at the base of a mountain slope; esp. such a deposit of fine-grained texture (silt or silty clay) deposited during time of flood.

Alluvial plain   A level or gently sloping tract or a slightly undulating land surface produced by extensive deposition of alluvium, usually adjacent to a river that periodically overflows its banks; it may be situated on a flood plain, a delta, or an alluvial fan.

Anastomosing   The branching and rejoining of channels to form a netlike pattern.

Avulsion   A sudden cutting off or separation of land by a flood or by an abrupt change in the course of a stream, as by a stream breaking through a meander or by a sudden changes in current, whereby the stream deserts its old path for a new one.

B horizon   A mineral horizon of a soil, below the A horizon, sometimes called the zone of accumulation and characterized by one or more of the following conditions: an illuvial accumulation of humus or silicate clay, iron, or aluminum; a residual accumulation of sesquioxides or silicate clays; darker, stronger, or redder coloring due to the presence of sesquioxides; a blockly or prismatic structure.

Bajada   A broad, continuous alluvial slope or gently inclined detrital surface, extending along and from the base of a mountain range out into and around an inland basin, formed by the lateral coalescence of a series of separate but confluent alluvial fans, and having an undulating character due to the convexities of the component fans; it occurs most commonly in semiarid and desert regions, as in the southwestern United States.   A bajada is a surface of deposition, as contrasted with a pediment (a surface of erosion that resembles a bajada in surface form), and its top often merges with a pediment.

Calcic horizon   A secondary calcium carbonate accumulation in the lower B-horizon that occurs as coatings on clasts and as lenses in fine-grained sediment matrices; it is at least 15 cm thick and contains 15 percent or more calcium carbonate.

Clast   An individual constituent, grain, or fragment of a sediment or rock, produced by the mechanical weathering (disintegration) of a larger rock mass.

Colluvial hollow   A bowl-shaped concavity in bedrock that collects sediment between debris flows.

Colluvium   (a) A general term applied to any loose, heterogeneous, and incoherent mass of soil material or rock fragments deposited chiefly by gravity-driven mass-wasting usually at the base of a steep slope or cliff; e.g. talus, cliff debris, and avalanche material. (b) Alluvium deposited by unconcentrated surface run-off or sheet erosion, usually at the base of a slope.

Crenulation   Small-scale folding that is superimposed on larger-scale folding. Crenulations may occur along the cleavage planes of a deformed rock.

Debris flow   A mass movement involving rapid flowage of debris of various kinds under various conditions; specifically, a high-density mudflow containing abundant coarse-grained materials and resulting almost invariably from an unusually heavy rain.

Delta   The low, nearly flat, alluvial tract of land deposited at or near the mouth of a river, commonly forming a triangular or fan-shaped plain of considerable area enclosed and crossed by many distributaries of the main river.

Dendritic   A tree-like pattern, typical of most drainage networks.

Desert pavement   Surfaces of tightly packed gravel that armor, as well as rest on, a thin layer of silt, presumably formed by weathering of the gravel. They have not experienced fluvial sedimentation for a long time, as shown by the thick varnish coating the pebbles, the pronounced weathering beneath the silt layer, and the striking smoothness of the surface, caused by obliteration of the original relief by downwasting into depressions.

Desert varnish   A dark coating (from 2 to 500 microns thick) that forms on rocks at and near the Earth's surface as a result of mineral precipitation and eolian influx.   The chemical composition of rock varnish typically is dominated by clay minerals and iron and/or manganese oxides and hydroxides, forming red and black varnishes, respectively. With time the thickness or the coating increases if abrasion and burial of the rock surface do not occur. As a result, clastic sediments on alluvial fan surfaces that have been abandoned for long periods of time have much darker and thicker coatings of varnish than do younger deposits.

Diffluence   A lateral branching or flowing apart of a glacier in its ablation area. This separation may result from the glacier's spilling over a preglacial divide or through a gap made by basal sapping of a cirque wall, or from downvalley blocking at the junction of a tributary glacier. Can be used to describe similar processes in water flow.

Eolian   Pertaining to the wind; esp. said of rocks, soils, and deposits (such as loess, dune sand, sand some volcanic tuffs) whose constituents were transported (blown) and laid down by atmospheric currents, or of landforms produced or eroded by the wind, or of sedimentary structures (such as ripple marks) made by the wind, or of geologic processes (such as erosion and deposition) accomplished by the wind.

Fluvial   Of or pertaining to or living in a stream or river; produced by river action, as in a fluvial plain.

Friable   Said of a rock or mineral that crumbles naturally or is easily broken, pulverized, or reduced to powder, such as a soft or poorly cemented sandstone. (b) Said of a soil consistency in which moist soil material crushes easily under gentle to moderate pressure (between thumb and forefinger) and coheres when pressed together.

Hydrographic apex   The highest point on an alluvial fan where flow is last confined.

Interfluve   The area between rivers; esp. the relatively undissected upland or ridge between two adjacent valleys containing streams flowing in the same general direction.

Lithology   The description of rocks, esp. sedimentary clastics and esp. in hand specimen and in outcrop, on the basis of such characteristics as color, structures, mineralogic composition, and grain size.

Loam   A rich, permeable soil composed of a friable mixture of relatively equal and moderate proportions of clay, silt, and sand particles, and usually containing organic matter (humus) with a minor amount of gravelly material. It has somewhat gritty feel yet is fairly smooth and slightly plastic. Loam may be of residual, fluvial, or eolian origin, and includes many loesses and many of the alluvial deposits of flood plains, alluvial fans, and deltas.

Loamy   Said of a soil (such as a clay loam and a loamy sand) whose texture and properties are intermediate between a coarse-textured or sandy soil and a fine-textured or clayey soil.

Morphology   The external structure form, and arrangement of rocks in relation to the development of landforms; the shape of the Earth's surface; geomorphology.

Pediment   A broad, flat or gently sloping, rock-floored erosion surface or plain of low relief, typically developed by subaerial agents (including running water) in an arid or semiarid region at the base of an abrupt and receding mountain front or plateau escarpment, and underlain by bedrock (occasionally by older alluvial deposits) that may be bare but more often partly mantled with a thin and discontinuous veneer of alluvium derived from the upland masses and in transit across the surface. The longitudinal profile of a pediment is normally slightly concave upward, and its outward form may resemble a *bajada* (which continues the forward inclination of a pediment).

Piedmont   (adj.) Lying or formed at the base of a mountain or mountain range; e.g. a *piedmont* terrace or a *piedmont* pediment. (n.) An area, plain, slope, glacier, or other feature at the base of a mountain; e.g. a foothill or a bajada. In the U.S., the Piedmont is a plateau extending from New Jersey to Alabama and lying east of the Appalachian Mountains.

Probability   The quantification of risk.

Relict   A landform that has survived decay or disintegration (such as an *erosion remnant*) or that has been left behind after the disappearance of the greater part of its substance (such as a *remnant island*).

Relief ratio   The average slope of a drainage basin; the ratio of maximum relief to basin length.

Rheology   The study of the deformation and flow of matter.

Riverine   Pertaining to or formed by a river. Situated or living along the banks of a river; e.g. a "riverine ore deposit."

Rock varnish   *See desert varnish.*

Scarp   (a) A line of cliffs produced by faulting or by erosion. The term is an abbreviated form of *escarpment*, and the two terms commonly have the same meaning, although "scarp" is more often applied to cliffs formed by faulting. (b) A relatively steep and straight, cliff-like face or slope of considerable linear extent, breaking the general continuity of the land by separating level or gently sloping surfaces lying at different levels, as along the margin of a plateau, mesa, terrace, or bench.

Schist   A strongly foliated crystalline rock formed by dynamic metamorphism which can be readily split into thin flakes or slabs due to the well-developed parallelism of more than 50 percent of the minerals present.

Scour   (a) The powerful and concentrating clearing and digging action of flowing air or water, esp. the downward erosion by stream water in sweeping away mud and silt on the outside curve of a bend, or during time of flood. (b) A place in a stream bed swept (scoured) by running water, generally leaving a gravel bottom.

Sheetflood   A broad expanse of moving, storm-borne water that spreads as a thin, continuous, relatively uniform film over a large area in an arid region and that is not concentrated into well defined channels; its distance of flow is short and its duration is measured in minutes or hours. Sheetfloods usually occur before runoff is sufficient to promote channel flow, or after a period of sudden and heavy rainfall.

Sheet flow An overland *flow* or downslope movement of water taking the form of a thin, continuous film over relatively smooth soil or rock surfaces and not concentrated into channels larger than rills.

Slurry A very wet, highly mobile, semiviscous mixture or suspension of finely divided, insoluble matter; e.g. a muddy lake-bottom deposit having the consistency of a thick soup.

Solifluction The slow (normally 0.5-5.0 cm/yr), viscous, downslope flow of waterlogged soil and other unsorted and saturated surficial material.

Stochastic hydrology That branch of hydrology involving the manipulation of statistical characteristics of hydrologic variables with the aim of solving hydrologic problems, using the stochastic properties of the events.

Stochastic process A process in which the dependent variable is random (so that the prediction of its values depends on a set of underlying probabilities) and the outcomes at any instant is not known with certainty.

Stratigraphy (a) The branch of geology that deals with the definition and description of major and minor natural divisions of rocks (mainly sedimentary, but not excluding igneous and metamorphic) available for study in outcrop or from subsurface, and with the interpretation of their significance in geologic history: It involves interpretation of features of rock strata in terms of their origin, occurrence, environment, thickness, lithology, composition, fossil content, age, history, paleogeographic conditions, relation to organic evolution, and relation to other geologic concepts. (b) The arrangement of strata, esp. as to geographic position and chronological order of sequence.

Swale (a) A slight depression, sometimes swampy, in the midst of generally level land. (b) A shallow depression in an undulating ground moraine due to uneven glacial deposition. (c) A long, narrow, generally shallow, trough-like depression between two beach ridges, and aligned roughly parallel to the coastline.

Tafoni Natural cavities in rocks formed by weathering.

Topographic apex The head or highest point on an active alluvial fan.

Translatory wave A gravity wave that propagates in an open channel and results in appreciable displacement of the water in a direction parallel to the flow.

Wash (a) A term applied in the western U.S. (esp. in the arid and semiarid regions of the south west) to the broad, shallow, gravelly or stony, normally dry bed of an intermittent or ephemeral stream, often situated at the bottom of a canyon; it is occasionally filled by a torrent of water. (b) Loose or eroded surface material (such as gravel, sand, silt) collected, transported, and deposited by running water, as on the lower slopes of a mountain range, esp. coarse alluvium.

List of Acronyms

BFE   Base Flood Elevation

CLOMR   Conditional Letter of Map Revision

FEMA   Federal Emergency Management Agency

FIRM   Flood Insurance Rate Map

HEC   Hydrologic Engineering Center

LOMA   Letter of Map Amendment

LOMR   Letter of Map Revision

NFIP   National Flood Insurance Program

NRC   National Research Council

NRCS   Natural Resources Conservation Service

SFHA   Special Flood Hazard Area

USACE   U.S. Army Corps of Engineers

USDA   U.S. Department of Agriculture

USGS   U.S. Geological Survey